Systems Analysis for
Business Data Processing

Other books by the same author

Data Processing Systems Design
Accounting and Computer Systems (co-author)

Systems Analysis
for
Business
Data Processing

H. D. CLIFTON, BSc. MBCS
PRINCIPAL LECTURER IN BUSINESS INFORMATION SYSTEMS
THE POLYTECHNIC, WOLVERHAMPTON, ENGLAND.

REVISED EDITION

petrocelli books, New York 1974

Copyright © 1974 Harold Dennis Clifton

Library of Congress Catalog Card Number: 75-124623
Standard Book Number: 87769-047-2

First U.S. edition in 1970
Revised U.S. edition, 1974 by
Petrocelli Books
div. Mason & Lipscomb Publishers, Inc.
384 Fifth Ave.
New York City 10018

First Printing

Printed in the United States of America

Library of Congress Cataloging in Publication Data

Clifton, Harold Dennis.
　　Systems analysis for business data processing.

　　Includes bibliographical references.
　　1. System analysis. 2. Electronic data processing
--Business. I. Title.
T57.6.C55　1974　　　　　658'.05'4　　　　74-3450
ISBN 0-88405-025-4

CONTENTS

PREFACE

The usage of computers, in all their variants, now permeates all aspects of commercial, industrial, and administrative procedures. No company or government department—local or national—is unaffected by this development; and significant numbers of these organizations are totally dependent upon the power of the computer for their continuing operation. The past decade has seen data transmission become amalgamated with computer-based data processing, with the result that the computer industry has become more correctly identified as the communications industry.

This revised edition is intended for prospective systems analysts and students of systems analysis and data processing. With the latter in mind, problems have been added at the end of each chapter together with suggested solutions. These are intended not only as intellectual exercises but also as supplementary information about the pertinent topics. The reader is therefore advised to study each problem and its solution, even if he is disinclined to attempt the problems for intellectual motives.

It is preferable for the reader to have had either a few years of experience of business methodology or some previous education in that field. Those persons coming completely fresh to the subject are advised to acquire a basic knowledge of accountancy and computer hardware and software before tackling this text.

Accountants and business managers wishing to understand data processing and systems analysis will find sufficient information to meet their needs; for these people, the more technical aspects in Chapters 5 to 8 can be omitted.

University and college students reading business studies and computer science will find that this book contributes considerable material to their courses. Preferably, this text should be studied in conjunction with the case studies from the companion volume, *Data Processing Systems Design.*

Computer programmers who wish to become systems analysts will have no difficulty in merging the information herein with their existing knowledge.

Every effort has been made to eliminate unnecessary technical jargon from the text, and increasing emphasis has been placed on financial and human considerations. The terminology employed is, for the most part, that already encountered by United States computer users.

H. D. CLIFTON

ix

CHAPTER 1

INTRODUCTION

1.1 MEANING AND SIGNIFICANCE OF DATA PROCESSING

An absolutely watertight definition of the term "data processing" is a somewhat ambitious and difficult aim in today's rapidly changing world. We can, however, regard data processing as being the means of obtaining control of business and administration through the provision of timely information of a high quality. Hence, as an alternative, we could refer to "Business Information Systems" because this name in many ways is more meaningful as applied to this subject.

Data can be regarded as being raw facts that are in some way recordable and from which useful inferences can be eventually drawn. Within the general framework of business, the word "data" pertains to the identification and measurement of objects, events, and people. Therefore, it is apparent that data is created continuously by all types of business activity and, in fact, by life in general.

The word "processing" refers to any series of actions and operations that enables useful information to be distilled and extracted from raw data. The implication here is that the latter is not in itself useful. This is true to the extent that raw data does not normally provide sufficient information to initiate direct action. When data is processed, the outcome is information upon which decisions regarding future activities and policies can be based.

What is it, then, that makes data processing so new or so different? When all is said and done, data has been around for a long time, and man has been using it to make inferences ever since his evolution as an intelligent thinker. Men also provided themselves with usable information long before the advent of electronic machines. Since accounts clerks, corporation presidents, factory

1

storemen, and corner shopkeepers have always used data, they are justified in asking, "What is special about data processing from my viewpoint?" These people, along with many others, have long been involved with processed data in the course of their everyday jobs. A corporation president is not in a position to present his firm's financial status until a very considerable amount of data has been processed. A shopkeeper, checking his bills, hardly sees himself in the role of data processor, but, as part of his job, that is sometimes his function.

We must avoid thinking of data processing as merely financial procedures; data can apply to the characteristics of events, persons, and things as authentically as it does to money. Data subjected to processing may be drawn from many spheres of activity, and it includes items such as road accident details, examination marks, share prices, and seat reservations. In the broad sense, data processing involves all basic facts from which useful information can result, and this applies in the business world especially to information that enables management to make far-reaching decisions.

The essence of data processing lies in the ability of modern computing systems to transmit and digest vast amounts of raw data at very high speed. The business methodologies now being employed were inconceivable before the days of electronic computation, magnetic storage, and high-quality telecommunications. This new power is based not so much on the computer's ability to perform complicated calculations as upon its capacity to organize data, i.e., to sort, store, and compare numbers and names. The storage devices now forming part of a computer configuration facilitate the rapid scanning and updating of large files of records, known as the system "data base." This means that more data than ever before can be taken into consideration when preparing information for management so that it can control the organization's day-to-day activities.

The compactness and accessibility of magnetically recorded files, combined with high-speed telecommunications, engender the centralization of the firm's records of all types within the data processing department. As a consequence of this, integration of the firm's planning and control activities becomes possible.

The creation of an integrated data processing system necessitates systems investigation and design of the highest order. This work is of a higher level than "organization and methods," involving not only a complete reappraisal of the organization's methods for achieving its objectives, but also the reassessment by top management of its information requirements for present and future control of the organization. Data processing systems are now being designed with a much more flexible nature to meet management requirements. When desirable, they can be arranged to allow managers to participate directly in obtaining selected information.

1.2 EVOLUTION OF DATA PROCESSING

Business organizations have had key-operated accounting machines of various types at their disposal for over half a century. These machines have been developed from simple mechanical adding devices into today's sophisticated electronic accounting machines and visible record "computers." Their means of output are now not only printed documents but also *paper tape* and *magnetically striped cards*. The latter give them the ability to record *carried-forward data* in a form that is easily re-input to the machine; this, together with an *internally stored program of instructions,* provides a limited degree of automatic operation.

A parallel but quite separate development was that of punched card machines. Punched cards, originally designed for special purposes such as the analysis of census statistics, were in use for over 40 years in business and government departments. Punched card machines were originally somewhat crude electromechanical devices, and in view of their large size, capable of doing surprisingly little in the way of business tasks. The mainstay of the system was the sorter; this machine, although very slow by modern sorting standards, was unsurpassed when it came to sorting large volumes of data.

When the printing tabulator appeared in the early 1930s, the first move had been made toward an automatic data processing machine. Never before had an accounting machine been able to run continuously without human control. Accompanying the printing tabulator came various machines for automatically arranging and punching cards. Examples were the collator for matching and merging groups of cards, the interpreter for printing the contents on the card itself (now done by card punches instead), the reproducer for copying the contents of cards into other cards, and the gang summary punch for automatically punching summarized results from the tabulator. Collators and reproducers are still in use to some extent as ancillary machines in data processing departments.

Not until the mid-1950s did punched card equipment make use of electronics, at which time its range was increased by the introduction of electronic calculators. These were able to do multiplication, division, and simple comparisons under the control of a wired program of instructions. Later, more advanced types of calculators were able to carry out quite complex calculations; and so much so that some were employed to do scientific calculations in the fields of atomic research and aircraft design. The employment of punched card machines in business organizations was generally only for the straightforward aspects of payroll, costing, stock control, and accounting—and then only for the larger companies and local authorities with sufficient volumes of data to justify the use of these machines in view of their restricted capabilities.

What, then, are the disadvantages of these two streams of equipment as compared with modern computers? The key-operated machines are geared inherently to the human operator. This has the advantage of versatility in that a wide range of tasks can be carried out with little time spent in preparation; but it has the disadvantages of comparatively low speed of throughput and of human fallibility in operation. The punched card machines, on the other hand, after being set up by the insertion of wired panels, were free from human error and were much faster in operation than the key-operated machines. They suffered, however, from the disadvantage of being inflexible, their functions being too rigid and narrow to allow for their use as management tools in a changeable situation.

A third and distinct development took place in the early 1950s in connection with scientific computing. Because of the tremendous increase in the complexity of scientifically designed plant and machines (such as atomic power stations, aircraft and weapon systems), the demand for computing power surged. This demand was largely met by the introduction of new models of computers, small by today's standards but of advanced concept and design for that period. Unfortunately these machines had limited printing capacity and this, together with the absence at that time of high-speed mass-storage media, meant that they were generally unsuitable as business computers, although a few were in fact used in this way.

The manufacturers of punched card machines also developed small computers based on the coupling of tabulators and gang punches to electronic units (e.g., IBM 650). In common with the scientific computers, these were controlled by internally stored instructions, and this feature gave them considerably more power than the punched card machines for planning and analysis purposes. Nevertheless, their speeds of input and output were geared to those of tabulators, and their only means of mass storage was punched card files. These limitations and the difficulty of programming made them unacceptable for most business applications. Another reason for their lack of success was the inability of both the business community and the manufacturers' salesmen to comprehend the potential of computers.

The first significant line of computers was introduced around 1960 by IBM, known as the 1400 series. During the years that followed, a large number of established punched card users were won over to this line, and of course this was facilitated by the compatibility of their existing cards. Similar computers followed from other manufacturers, and these so-called second-generation computers possessed three main advantages over their predecessors: first, higher data input/output speeds; second, means of storing mass data and processing it at high speed (i.e., magnetic tape); and third, higher-level languages to ease the programming burden.

The mid-1960s saw the introduction of the so-called third-generation computers, and among the present leaders of this field are the ICL 1900 series and the IBM system 360-370. The principal characteristics of third-generation computers that distinguish them from their forerunners are extensibility and compatibility. Extensibility, as the name suggests, means that the user can make his computer grow by replacing or adding to its units (known as hardware). A small computer can thus be gradually transformed into a large one without at any one time replacing the computer as a whole, thereby minimizing the disturbance of changing over. Compatibility implies the ability of a given range of variously sized computers to accept both each other's programs and those from a standard range (software). Also, any selection from existing or future peripherals (hardware) can be attached at will.

During the late 1960s, the principal developments were in the field of peripheral equipment, arising from the compatibility aspect of the third-generation computers. A considerable number of companies, some quite small, entered the data processing market with a wide range of peripheral units such as disk storage devices, visual display units, and on-line input keyboards; also available is microfilm equipment for use with computers. Thus, although the mainframe suppliers have been reduced in number owing to takeovers and mergers, the overall number of firms in the data processing business has increased considerably.

1.3 EMPLOYMENT OF COMPUTERS IN BUSINESS

Computers are nowadays an accepted part of business and governmental machinery; this is evidenced by the disappearance from the press of colorful accounts of computer applications and their replacement by discussions of the sociological problems brought about by computers. The amount of mundane but necessary work being done by computers is far greater than is generally realized, and in many cases it would be impossible to return to manual methods. Clerical work, in the old-fashioned sense, is a dying occupation but the replacement of men will not be by machines but by different kinds of men—be they the original men in other guises or a newer breed. It is the particular job, not the man, that is at stake. Tomorrow's clerk will have at his disposal a million-dollar computer, which he will command but probably never see. The complexity of modern planning and the rate of change of many business situations are so great that they cannot really be coped with except by means of a man-machine relationship. Provided men can adjust their attitudes so as to assert their intelligence superiority over the computer, this relationship will be harmonious and rewarding.

Business Applications and Recent Developments

Computers are currently in use in very many companies and organizations; their application runs the gamut of conventional business and administrative activities. It is interesting to identify a few examples from the extremely wide range of existing applications; from them, one can perceive the computer's versatility and potential.

Road accident recording and analysis.
Registration of shares.
Electronic injury surveillance.
Airline seat reservation.
Passenger transport surveys.
Blood donor administration.
Selection of shipping facilities.
Land-use surveys.
Tote betting accounting.
Election result forecasting.
Examination administration.
Portfolio evaluation.
Crime recording and statistics.
Retrieval of legal information.
Parking ticket accounting.

The preceding list is by no means comprehensive, but it is sufficient to demonstrate that a considerable amount of talent has been employed in making these applications realizable on computers. There was a considerable degree of fortuity in their implementation, their selection having arisen mainly because of the vision of the first user of the application. Now data processing power has become more readily available, and people are accustomed to having it on tap, the fortuitous and visionary aspects have disappeared. Developments have removed the barriers between the "ordinary" man and the computer; a brief mention of several developments in this direction is worthwhile at this point.

The bigger the computer, the better the value for money. It is therefore generally more economic to use a small amount of time on a big computer than a large amount of time on a small computer. The former aim can be achieved in one of two ways—either by renting time on a large computer and using the machine fully during that time, or by "time sharing" simultaneously with other users. Time sharing involves the installation of terminal hardware and the provision of data transmission facilities, but it has an advantage over renting time in that the exact time of usage does not normally have to be preallocated.

Another aspect of man/computer communication is the formation of modes of expression common to both. This is seen in an elementary form in the employment of stylized printing on documents; this is recognizable by machines and men, and is known as optical character recognition (OCR), or magnetic ink character recognition (MICR).

A more sophisticated form of common expression is the system known as "conversational mode."* This brings the human and electronic counterparts into contact by permitting a pseudo-conversation to take place between them. The computer is programmed to respond to the user in such a way as to make it appear that it understands his instructions. At the same time, the computer is able to lead the user along the correct path toward obtaining his requirements. A simple version of conversational mode is the use of "computer utilities"; with these, the user is in contact with the computer via a keyboard and a printer or visual display unit at his terminal. The job in hand is preplanned to be done by an interplay of the user and the computer, thus giving a degree of flexibility more akin to keyboard machines than conventional computers, but at the same time preserving the computer's calculating and organizing power.

Allied with the developments mentioned above is the increased employment of visual methods and, to a lesser degree, audio input and output. The main visual technology is the cathode-ray unit, arranged to provide a display of figures or graphs and able to accept modifications to the data by means of a light pen or similar device. The business application of this equipment has resulted in a much closer contact between the manager and his files, the rapid speed of display enabling large amounts of data to be shown without delay, and thus facilitating the interrogation of files from the manager's desk. Similarly, the input of data such as customer's orders is brought closer to its source and is also brought into closer contact with the master files. These features provide for more stringent vetting of the input data and easier subsequent correction of the errors thus detected.

Why Analyze Systems?

An inspection of the short list of less well-known applications given previously reveals the fact that they are mostly isolated or occasional applications. For these specialized applications this is generally of no consequence, but even with more conventional computer applications there is an underlying tendency for this to be the case. It is far easier to build a house in the country than to rebuild a town; similarly, companies that have successfully implemented isolated computer jobs have failed when attempting to integrate them to form a total system. Without stretching the analogy too far, a

*Martin, *Design of Man-Computer Dialogues,* Prentice-Hall, 1973.

town can arise from uncoordinated construction projects, but only after makeshift alterations can it be made to function as an entity.

The fragmented implementation of computer work can cause similar problems; it is dangerous to allow each department to specify its requirements in isolation and then to design a system to provide these quite separately. The requirements of department A may conflict with those of department B; alternatively they may coincide, but in any event it is better that they are considered together. Only through a painstaking investigation and analysis of the overall situation within the organization can a comprehensive plan evolve.

This is not to suggest that all applications must be implemented simultaneously, nor that it is absolutely imperative that every application is invariably included in the data processing system. The main objectives of systems analysis are (1) to study in depth the aims and problems of existing work, and (2) to design a system that is "open ended" so that further applications can be welded to it without duplication of work or records.

CHAPTER 2

THE SYSTEMS ANALYST

2.1 THE NEED FOR FULL-TIME SYSTEMS ANALYSTS

In the preceding chapter the usages of hardware (machines and equipment) and software (the means of controlling machines and systems) have been mentioned briefly. We now turn to the employment of people in the world of data processing. Because of the considerable publicity given to computers and their mathematical capabilities, there has developed a tendency to overlook the people associated with them. Hard experience has shown that in many cases this has led to inefficient and fragmentary employment of expensive computing equipment.

Owing partly to this state of affairs, the occupation of "systems analyst," as such, has evolved. His work was done previously by various other personnel, but within a narrower framework. These people were involved on a part-time basis and, more often than not, with isolated applications. Their efforts were frequently frustrated by lack of time and by insufficient training in data processing methods.

When the first computers were installed, it was immediately recognized that programmers were necessary, and each user employed a few bright young men in this capacity. Early computer applications, such as payroll and sales analysis, were implemented successfully due to their inherent definability. Later applications, for instance production control, often ran into difficulties, owing to the programmers being unable to formulate the work into neat computer routines. During the attempts to implement later applications, it became apparent that a communication barrier existed between programmers and management. The programmer's attitudes were dominated by the intricacies of instructing the computer. Management, on the other hand, which

9

was adept at controlling men, had a blind spot as regards the meticulous detail required for computer utilization.

This situation virtually halted the transference of further work to the computer, and so it became used merely as a sophisticated accounting machine. The employment of a computer for integrated applications is not possible without a thorough investigation of the organization and an overall replanning of the existing systems.

How can these tasks be achieved? Only by employing a full-time systems staff, recruited from within or outside the company, and thus creating an environment for systems analysis in its own right. Allocation of staff from other departments and part-time work in this capacity are very unsatisfactory solutions. Naturally such staff show loyalty to their own departments, resulting in bias in favor of that work. At the same time, the personnel of other departments are suspicious and, when their department is investigated, fear that a takeover bid is being made.

2.2 THE FUNCTION AND DUTIES OF THE SYSTEMS ANALYST

Although the systems analyst can be regarded generally as an "agent of change," his precise duties and responsibilities vary from one company to another. He is being increasingly employed by organizations representing the whole spectrum of the industrial, business, and administrative societies. His employers, in addition to the usual range of manufacturing and commercial companies, include consultant firms, service bureaus, computer suppliers, educational institutions, and (by no means least) government, both local and national. It is therefore obvious that the range of his tasks is diverse; nevertheless the broad principles guiding his approach are the same for all situations.

The title "Systems Analyst," although slightly misleading, has by now been assimilated into everyday language in connection with the employment of computers for data processing. The vast majority of systems analysts are concerned with business applications of computers; it is solely in this aspect, as opposed to scientific problems, that we are interested herein.

The systems analyst could equally well be entitled "Systems Investigator," "Systems Designer," or "Systems Implementer," since all these aspects are part of his function.

His range of activities consists of:

1. The investigation and recording of existing business systems with a view to discovering inefficiencies, problems, and bottlenecks.

2. The analysis of data acquired during the investigation so as to prepare it for subsequent utilization in a data processing system.
3. The design and appraisal of new systems, usually involving computers, bearing in mind the objectives set by management.
4. Assisting with the implementation, documentation, and maintenance of the new system.

Systems Project Teams

As stated earlier, the precise nature of the systems analyst's activities varies between organizations. In a large company the analysts are normally organized into project teams, each team concentrating on one or more applications. The more senior analysts act as team leaders so as to coordinate and guide the work of the less experienced members of the team. Where teams exist, it is essential that their individual efforts are coordinated at every stage, and not merely forced together after completion. This can be accomplished by the systems manager's arranging regular meetings between teams, and by his encouraging continuous intercommunication between them.

Instances of the need for coordination between project teams occur in a situation where the total work is split into, say, three parts. Project team A might be assigned to the accounting functions of the company; team B, to production control; and team C, to payroll. The teams obviously have many points of contact in their respective work. In particular, teams A and B interact through the control of sales and purchase orders; teams B and C, through factory payment and labor costing; teams A and C, through control of cash and bank accounts.

Programmer/Analyst

Within some smaller companies, and indeed in certain larger organizations, the same person does the work of systems analysis and of program writing. The efficacy of employing systems programmers/analysts (not to be confused with systems programmers, who do a different job) depends largely upon the personal qualities of the individuals concerned. Provided the programmer/analyst is of sufficiently high caliber, both socially and intellectually, to contend with the dual role, there are several advantages to be gained. The foremost of these is the elimination of the communications problem that can exist between the systems analyst and the programmer (Section 2.6). Another significant advantage is with regard to the flexibility of staff disposition; this is increased because each man can do either job. This versatility results in a smoother work load, and in a reduction of the idle time that can be caused through the staff's awaiting the output of work from each other. The main problem connected with the dual role is caused by the clash in the attributes desirable for each occupation. The systems analyst must be fairly extroverted and be content to spend many hours talking to people in their "language." The programmer, on the other hand, must be prepared to spend even more hours working individually with the computer's language. It is unusual for these different traits to be possessed by one and the same person.

2.3 DESIRABLE KNOWLEDGE AND ATTRIBUTES

A list of the desirable qualities for a systems analyst reads as a testimonial for perfection in both intellect and attitude. It engenders in its reader a skepticism toward belief in the existence of such a talented person. Nonetheless, these desirable qualities are worth reflecting upon and, if borne in mind, will help the prospective analyst to achieve his aims.

Education

A systems analyst's formal education should be equivalent to at least a good final level at junior college, and although higher academic qualifications are not strictly necessary, an increasing proportion are in fact graduates in one or another of a wide range of subjects. It is probable that the majority of systems analysts in the future will have academic qualifications in business studies, business information systems, or computer science prior to their recruitment into this occupation.

Background Experience

Between the dates of qualifying educationally and starting work as a systems analyst, it is very desirable that the applicant's background experience will have been widened. Several years spent in a variety of companies or departments provide invaluable experience for this occupation, as for many others. Ideally, this background experience embraces topics such as accounting, costing, planning, and control. An extensive background of computer programming is of dubious advantage and, on its own, is not an acceptable qualification. In the past, results have shown that only about one in three good programmers has subsequently made a good systems analyst. In a similar way, a long period spent in one department only is likely to have instilled a narrow outlook and to have made the person hidebound by his previous experience.

Knowledge

The extent and pattern of the knowledge needed by the systems analyst depends largely upon his precise duties and upon the type of organization for which he is working. His knowledge need not necessarily include mathematics, but a comprehensive understanding of arithmetic and the ability to maneuver numbers are very desirable. Nor does it include formal qualifications in accountancy, but a good general knowledge of this subject is very useful.

Further desirable knowledge would include:

1. A broad understanding of business procedures.
2. The routines and techniques involved in all or some of the following:

   Production planning and control.
   Stores and stock control.
   Accounting procedures.
   Administration, personnel and general.
   Marketing and purchasing.
   Preparation of surveys and analyses.
   Operational research in business.

3. An understanding of the purpose and objectives of the organization from the viewpoint of its top management.
4. The techniques of systems analysis and data processing, including some familiarity with programming strategy and languages.
5. A comprehension of the range of data processing equipment (hardware) that is currently available and the systems software associated with it, especially business application packages.

Attributes

The following attributes will prove to be useful in systems work even though few persons can be expected to have them all.

1. A capacity to absorb quickly the wide concept of a system and also to give meticulous attention to its detail.
2. The ability to get along with all types of people and particularly with all levels of management, clerical and administrative staff, and programmers—to some degree the analyst must be "all things to all men."
3. The possession of a self-assured but not flamboyant manner.

4. A scientific skepticism, being interested in, but critical of, existing methods and information given to him. The ability to adopt a detached attitude is a valuable asset when interdepartmental discussions become heated.

5. Patience—a willingness to explain new methods, repeatedly if necessary, to other members of staff, including the programmers. Also patience in waiting for people to collect facts and for their ideas to mature.

6. A genuine desire to explore situations in order to discover the underlying factors, and an eagerness to learn new methods, techniques, and developments in hardware.

7. A capacity to think logically combined with a "fail-safe" mentality—this facilitates systems design, especially from the security aspect.

8. The ability to communicate verbally and in writing, and thereby transmit enthusiasm for new ideas and information about new systems.

9. Perception combined with tact, so as to be a good prober while not giving the impression of grilling people.

10. A willingness to admit that certain problems extend beyond his knowledge and, accordingly, to summon specialist assistance in such areas as operations research, statistics, and econometrics.

11. The ability to make good notes, combined with a reasonably good memory. An exceptional memory is not necessary; the amount of facts involved would swamp the best of memories anyway.

12. Diplomacy—so as not to become regarded as the "hatchet" man or the "efficiency" man.

2.4 RELATIONSHIP WITH MANAGEMENT

The relationship between the systems analyst and top management depends largely upon the size and structure of the organization. In a large company it is obvious that every systems analyst cannot have unrestricted access to the managing director. In a small company, however, this situation may be both possible and desirable. Hard and fast rules cannot be laid down to cover all circumstances, but whatever prevails, a firm link must be forged between the planners and the decision makers. This link between analysts and management, if not direct, is via the data processing manager or the management services manager. Whosoever is involved as intermediary must be capable of appreciating the significance of the analyst's work so that he can support the resultant recommendations during top-level discussions. The link should be as short as is possible without invalidating the organizational structure of the company.

During the early stages of a system's investigation, a different situation may exist. At this point in time it is unlikely that management services and data processing will have been established. Their existence will probably be somewhat nebulous, and consequently there is reason for more direct contact between the systems analyst and top management. In some ways, the caliber of the analyst employed at the outset must be higher than that of later comers. Above all, he must be capable of talking to top managers in their own terms and must understand their problems. His attitude should be such as to convince management that he is far more concerned with the advantages that data processing can bring to the organization than with computers in themselves. Unfortunately, this has not always been the state of affairs in the past. As a consequence, certain systems analysts have become regarded by management as being out of touch with the harsh realities of business. A systems analyst, and particularly one with a long background of data processing and computers, must always remember that as far as management is concerned, the computer is merely another tool in the firm—and quite rightly so. A computer does not merit an accolade simply because of its existence, but only when its value to the firm has been proved over a long period of time.

Another attitude that the systems analyst should avoid adopting, perhaps subconsciously, is that of siding automatically with computer manufacturers. It is, of course, even worse if he has a strong unreasoned bias toward one particular manufacturer or against another. We are all conditioned to some extent by our past experience, especially as regards personal factors, but the systems analyst must rid himself of any emotional bias and operate solely on objective factors. He is a paid employee of the firm for which he works; as such, he owes his allegiance entirely to that firm. This must be completely evident to his management, and unless this is so, he will not be fully accepted and trusted by his employers.

The situation in a smaller company or division is somewhat less constrictive, for there is usually the opportunity here to develop a close rapport between systems and management. Provided the systems analyst does not push his recommendations too hard and is able to demonstrate that he understands the problems as well as the potential advantages of a new system, his success is better assured. The mutual empathy fostered between the two parties can do nothing but good, not only for the data processing system but also for the systems analyst's personal career.

2.5 COOPERATION WITH OTHER DEPARTMENTS

During the course of his investigations, the systems analyst may find himself in the paradoxical position of requiring the utmost cooperation from some of the departments in order to phase out their outmoded operations. Whether or not this is the case, the advice of departmental staffs has no substitute, and must be sought with diligence. The sum of their experience accumulated over the years provides the distillates for a new system. By realizing this, and accepting that he is not an expert on every subject, the analyst has taken a step on the road to successful systems design.

When discussing the work with departmental managers, the systems analyst's remarks should be directed toward making them feel that they will form an essential part of the new system. Before asking them for their additional information requirements and suggestions for improvements, it is advisable for the analyst to explain briefly the potentialities of a computer system. This puts the data processing picture in perspective and encourages the managers to discuss information requirements previously considered to be unattainable. It may take some time for ideas to mature, and so the analyst should allow time for deliberation.

While giving full consideration to all advice that he receives, the systems analyst should guard against adopting outdated ideas and methods. This caution applies especially to systems created by the administration department and the punched card department in the days before computers arrived on the scene. Punched card methodology had, ostensibly, several similarities to modern data processing. In particular, these were the cards themselves, the punching thereof, and also the general philosophy of processing data in batches. Well-organized sets of code numbers are also likely to be bequeathed by punched card departments, and in all probability these will be suitable for computer use. Although the analyst may accept the advantageous features of punched card systems, it should be remembered that they often obviated difficulties and exceptions by interjecting manual procedures at certain points in the system. This is not usually acceptable in a computer-based data processing system, and so these problems must be dealt with in more elegant ways. A typical example of this sort of thing is inventory control. Whereas stock reports stemming from punched card systems were inspected manually for obsolescent items, these items can be detected and dealt with automatically by incorporating a usage analysis into the data processing system.

The Accounts Department

A rather special relationship is likely to develop between data processing and accounting,* and in quite a few cases the data processing system will consist primarily of accounting applications. It is therefore likely that the

---

*Clifton and Lucey, *Accounting and Computer Systems,* Petrocelli Books, New York, 1974.

company's chief accountant will become closely involved with data process-ing in several ways:

1. Initially, in decisions on the financial case for computer-based process-ing.
2. Liaison with systems personnel on systems investigation and design.
3. Monitoring and auditing the costs and benefits deriving from the data processing system.

By understanding the essential features of computer-based data process-ing, the accountant is able to recognize the potentialities of the computer and to anticipate difficulties arising during the design and life of the system. Although it is not possible to create a precise demarcation line or an exact interface between the responsibilities of the accountant and the systems analyst, this is a quite satisfactory situation, provided there is close discus-sion of the problems involved. Discussion not only insures that there is no gap between responsibilities, but also creates mutual understanding and appre-ciation of each other's knowledge and skills.

It should be remembered that the basic duties and responsibilities of the accountant are not changed by the adoption of data processing methodology. He has the right, as have other departmental managers, to decide the aims of data processing in relation to his department. As a user of the data processing department, he is in effect its "customer" and therefore has to decide what to "purchase" from it. The systems analyst ought not to "sell" him a system merely because it is prepackaged and thus easy to implement. The "com-modity" sold and bought must always be what is really needed.

2.6 HIS POSITION WITHIN THE DATA PROCESSING DEPARTMENT

In a data processing department with a total staff of 20 or more (excluding punch operators), it is likely that systems analysts will comprise one or more teams under the leadership of a senior analyst, who also acts as deputy data processing manager. Another arrangement is for the data processing manager to be in charge of programming and operating only, while the analysts form a separate department under the authority of a systems man-ager. This person reports directly to the management services manager or someone of similar status.

Whatever the arrangement, it is essential that the "interface" between the systems analysts and the programmers is clearly established at the outset.

Failure to do this can result in demarcation disputes arising during the programming and implementation phases.

What is meant by "interface" in this context?

1. The precise level of information to be provided by the systems analyst, and the form in which it is to be presented to the programmer (Sections 10.5 and 10.6).
2. The authority of the systems analyst in relation to disputes about the precise methods and techniques to be adopted. Programming problems must not be allowed to dominate the new system; its aims are paramount and therefore programming problems must be overcome either by adopting different programming methodology or by employing more skilled programmers.
3. The maintenance of contact between the systems analyst and the programmer during the programming phase. Contact should be close enough to insure that the programmer keeps "on target" and that no deviations creep into the program due to misunderstandings of the systems analyst's intentions. This can easily happen in systems routines involving complex decisions and consequent variations in the computer program. It may well be necessary to adopt precise predefined formats such as decision tables (Section 11.1) in order to eliminate the possibility of error.
4. The degree to which programmers are permitted to contact staff outside the data processing department for additional information, and the extent to which they are allowed to make amendments to the routines as specified by the systems analyst. Within limits, these tasks can be done by the more senior programmers, and there are two main advantages to this. First, in a situation where the analysts are hard pressed on further systems, they do not have to break off to deal with minor queries or changes. Second, it makes a natural introduction to systems work for programmers.

Although this interface should be established initially in a formal manner, it can be modified later when the parties have become more skilled and familiar with each other's capabilities. It is, however, advisable to preserve the formal definition of interface for the benefit of new data processing staff.

Promotion Prospects

Within an established data processing department, the systems analyst's line of promotion is normally via senior analyst to data processing manager or systems manager. Where the department has not yet been created, the systems analyst works alone and may become data processing manager designate if and when the need for data processing becomes apparent. Promotion to the position of computer operating manager would not suit the majority of analysts—this job is too routine for the explorative mind. In the larger organizations, a line of further promotion is to management services manager, and from here his wide knowledge of the organization makes him eligible for the highest management positions.

There is likely to be a continuing demand for experienced systems analysts from a wide range of industrial and administrative concerns. This, together with the unique opportunities for gaining experience available to the analyst, give him promotion prospects surpassed by few other occupations.

2.7 EXERCISES

Problem 1. Systems Staff Selection

The AMTO Company is a medium-sized firm manufacturing machine tool equipment. It owns a computer that is currently in use for carrying out basic accounting work. One experienced systems analyst and three programmers are employed on data processing work, and it is intended to supplement this staff by the addition of one more systems analyst. This person will be responsible primarily for investigating and designing integrated computer-based inventory control, production control, and allied areas. When once established in the post, he will not receive much assistance from the existing systems analyst because this person is already fully occupied with other tasks. He may, however, be allowed some junior assistants at a later date if this is found to be necessary.

The following candidates apply for the post. Comment on their suitabilities, and suggest your preferences.

MR. ADAMS:
> Age 28, at school until 16, office clerk until 23, including two years in the Army (corporal in stores). During past five years he has been a programmer working on a variety of applications in a food distribution company and in local government. Moderately intelligent, keen programmer, careful documenter, rather overawed by management and of a reserved nature, but well accepted by his colleagues.

MR. BENTLEY:
> Age 25, business college until 20, insurance salesman until 24, now manager of a small insurance office. Ambitious, dynamic, hard working, but with no practical knowledge of either manufacturing industry or data processing, very extroverted, and inclined to exaggerate.

MISS CLAY:

> Age 30, business college to 21, bank clerk until 25, thereafter administrative work in an automobile factory. Experienced in applying punched cards and accounting machines, self-assured, and a good mixer.

MR. DEAN:

> Age 35, qualified and experienced accountant (13 years) in a variety of industries. Wishes for a change of occupation, but with opportunity to utilize his knowledge. Rather self-important and sees himself very much as a manager; some knowledge of data processing systems.

MR. EDWARDS:

> Age 22, fresh from university with a good economics degree, very intelligent, ambitious, and keen to get a good job. No practical experience of industry whatsoever; has taught himself COBOL and has read books on systems analysis.

Solution to Problem 1. The situation at AMTO calls for a person who is self-sufficient and who can work with management in developing the new systems. The need to computerize inventory and production control means that a candidate with industrial experience will be at an advantage, but the integration with accounting must not be overlooked.

Mr. Adams is not likely to be happy doing pioneering work that calls for frequent contact with management and user staff. Five years as a keen programmer tends to indicate that this is his forte. A possibility is that he might be considered for junior membership of a systems team at a later date. Provided his programming experience has given him some depth of understanding of the applications he has encountered, he should qualify as a junior systems analyst.

Mr. Bentley's obvious verve would make him a useful systems team member in the later stages of systems work. He could no doubt "sell" the system and assist in implementing it, but it is highly unlikely that he could design it. In this latter respect his lack of industrial and data processing knowledge would be a severe handicap.

Miss Clay has a good background for this post—her experience as an administration practitioner in an automobile factory implies that she is used to talking with production staff and that she is familiar with industrial situations. She would, of course, need some training in computer-based data processing, but this should cause no difficulty in view of her machine background. Provided no sociological problems may be caused by the employment of women at AMTO, Miss Clay would be very suitable.

Mr. Dean could undoubtedly cope with the systems planning because his extensive accounting experience is very likely to have provided him with the necessary knowledge of production and stock control. His drawback is his attitude; for a more senior post, this might be acceptable, but for this particular post it might make life difficult for his colleagues.

Mr. Edwards, in spite of his rawness, might be successful in this job. He would obviously go "all out" to learn the methodologies involved, both as regards industrial production and data processing. The big snag to this is that there is no time at AMTO for on-the-job learning, and consequently his inexperience must count against him.

To summarize these points, we must first remember that there is no absolutely correct answer—either on paper or in reality. The final choice of candidate is to some extent governed by subjective factors such as personality and temperament, and these cannot really be put down on paper. Nevertheless, based on what we do know about the candidates, the choice appears to be in this order:

1st—Miss Clay: well ahead of the others.
2d—Mr. Dean: in spite of his social attitude.
3d—Mr. Edwards: because of his enthusiasm and intellect.
4th—Mr. Adams: not really suitable but just acceptable.
5th—Mr. Bentley: unsuitable.

Problem 2. Systems Staff Selection

BUCOB inc., a computer service bureau company specializing in business applications, requires a systems analyst to both sell and plan data processing systems. This post is one of a number at BUCOB; consequently, (1) there is plenty of help available in case of difficulty, and (2) the successful candidate must be prepared to change rapidly from one job to another. The analysts' salaries consist partially of commission, and this is based upon the revenue earned by BUCOB from the work obtained and planned by the systems analyst concerned. A very wide variety of work is encountered by BUCOB's staff, and this is often obtained against stiff opposition from other bureaus. The jobs carried out vary considerably in both size and nature, but, generally speaking, they are self-contained and of a clear-cut type.

The same candidates as in Problem 1 apply for the post. Again comment on their suitabilities and suggest your preferences.

Solution to Problem 2. The BUCOB post calls for the ability to sell and to plan. These attributes do not combine too readily but nonetheless the candidate must be prepared to accept them. Ideally, broad experience combined with initiative and flair is required. In addition, the successful candidate will need to have the astuteness to detect profitable work, since his salary depends partly upon this.

Mr. Adams' knowledge of computer applications would serve him quite well, but this advantage is outweighed by the need to sell. He is likely to become very involved in the detail of a problem; this tendency, combined with his innate reserve, mitigates against his obtaining many orders.

Mr. Bentley is an obvious choice from the selling aspect. If given time to acquire data processing knowledge and experience, he would very likely be successful. If BUCOB is prepared to train him and allow him to work along with an experienced systems analyst for a year or so, Mr. Bentley will be a beneficial employee.

Miss Clay is a good candidate for the post. She has a fair amount of practical experience in industry, and her ability to mix with people, combined with her self-assurance, should stand her in good stead.

Mr. Dean would find the selling aspect of this post to be tough going in view of his self-important and managerial attitude. Counterbalancing this defect is his vast experience, and if he could be employed as a second-line systems analyst to follow the salesman, he would be very successful. This is not, however, the existing situation at BUCOB.

Mr. Edwards, being young and energetic, would be reasonably successful right from the start. His lack of experience could be mitigated by his careful selection of contracts gained until sufficient knowledge had been acquired. The snag with Mr. Edwards is that unless promotion was in sight, he would probably move on to a more prestigious occupation after a few years; this may not matter to BUCOB, however.

Summarizing the facts given above and bearing in mind the general comments in Problem 1, the order of preference might be:

1st—Miss Clay: just ahead.

2d—Mr. Edwards: based again on his enthusiasm and brains.

3d—Mr. Bentley: moderately suitable.

4th—Mr. Adams: not suitable.

5th—Mr. Dean: not suitable.

CHAPTER 3

THE OBJECTIVES
OF THE SYSTEM

3.1 IMMEDIATE AND LONG-TERM OBJECTIVES

An analysis of the replies to the question, "What are your immediate and long-term objectives?" would provide interesting if not altogether productive results. As might be expected, the replies would show a marked degree of correlation with the positions of the persons replying. Also, they would contain a proportion of irrelevant information that may obscure the real objectives. The chairman of a company may have the immediate objective of satisfying the shareholders at the next meeting, and the long-term objective of taking over a competitive firm. The production manager's objectives may be to reduce overtime working in the short term and to re-equip the machine shop in the long term. Whereas most members of staff will subscribe to the aim of profit making or of meeting a predetermined target of some sort, this is by no means their most overriding consideration. Quite naturally their attitudes are colored by personal ambitions, capabilities, and anxieties, resulting in distorted objectives. The systems analyst, particularly if he is new to the organization, will not always find it easy to dissect and evaluate these diverse and conflicting aims.

It is generally true that the higher a person's status in the organization, the longer the term of his objectives. It is also generally true to say that the long-term objectives form the true basis for systems planning, although short-term objectives often tend to dominate the situation. This can lead to an unhappy state of affairs if the two are in serious conflict, since it is very often difficult to eliminate the effects of short-term remedies. This is not to say that a computer is necessarily always more expensive than manual calculators, but that it is imperative to consider the longer term costs, taking all factors into account. For instance, regardless of cost, a computer is installed to counteract the effects on business caused by labor shortage. In the short term, the computer solves the problem. As a result, management adopts the objective of decreasing labor costs, based on this immediate solution. But then a surplus of labor becomes available so that over the long term the projected savings will not be realized because the computer cost exceeds the hiring price of the available labor. In this particular example, the purchase might still be justified because:

1. Other work could also be carried out.
2. There was no guarantee that the shortage would be temporary.
3. More accurate and comprehensive results would accrue to business benefits.

The systems analyst's major problem is to balance the immediate and long-term objectives when designing a new system. All aims and requirements must therefore be weighed—qualitatively if not quantitatively—before using them as the basis for a data processing system.

Financial Objectives

Virtually every organization is controlled by financial considerations, whether they are of profit, cost, or budget. In the early stages of a systems investigation, the high-cost areas should be ascertained so that they can be given considerable attention. The justification for a new system often relies upon a reduction in overall cost. In the past, this reduction has not always taken place, particularly with computer systems, and has resulted in some disillusionment on the part of top management. Why was this so? What went wrong?

Obviously there is no single ubiquitous reason, but a number of factors are prominent:

1. The new system was not comprehensive enough to completely replace the old system. Although the bulk of the routine work was taken over, often there were several fringe jobs that were not included. These were the awkward jobs involving intuitive human decisions, communication with other organizations, or other nonroutine tasks. In the desire to implement the new system speedily, these jobs were ignored, with the result that it was found later that more staff and machines were required than was originally visualized. In the worst situations, new systems have been gradually eroded by attempts to deal with unforeseen exceptions and emergencies. If a proposed system is justified by potential saving in costs, this must be made realizable

even if, for example, it entails a change in the level of customer service. It must be remembered that every service has an ascribable cost and value, however approximate. This must be taken into account along with other costs and savings, and if necessary, management must be prepared to make policy decisions regarding the future level of service and similar considerations.

2. Confidence in the new system was lacking, owing to its tendency to produce errors in results or incur delays in providing the necessary outputs. As might be expected, management has in the past tended to counterbalance these difficulties by keeping the original staff in reserve. The solution to this problem is found in reliable machines, good systems design, and meticulously careful programming. Every contingency should be covered by prescribing standby arrangements in case of machine trouble and by employing within the system every possible means of error detection.

3. The computer was overloaded because of an underestimation of the time taken by it to do the required work. The fantastic speeds that are theoretically achievable have misled planners into using them to estimate times in practical situations. This is somewhat analogous to purchasing a 120-mph car and expecting to travel from New York to Philadelphia in 55 minutes. The solution to this problem is to estimate more realistic timing of runs (see Section 11.2).

Other Objectives

Although truly nonfinancial objectives are very rare, it is convenient to group together several objectives that are ostensibly nonfinancial.

Alleviation of Staff Shortages. This includes not only the situation where there is a more or less permanent shortage, but also the one in which there is a regular turnover of staff. In the latter case, a department may always consist of a substantial proportion of newcomers and prospective leavers. This situation usually leads to inefficiency and consequent deterioration of staff morale, which in turn lead to more turnover. Other reasons for staff shortage are, for example, the boring nature of the work, the isolated location of the factory, the low rates of pay, and the poor social atmosphere in the department. If faced with these situations, a systems analyst must not assume that a new system will automatically cure all ills. The true nature of the trouble should be diagnosed before a remedy is recommended. For example:

1. *To cope with increased work:* The increase may be permanent or temporary, significant or slight, steady or fluctuating. In order to be sure which of these characteristics apply, it is necessary to discover the genuine

reason(s) for the increase. In many cases the reason is obvious, but its veri-fication can do no harm however. Each of the situations mentioned above can then be dealt with by adopting the appropriate measures, including the data processing facilities described in Section 12.3.

2. *To provide information for management:* This does not inevitably mean more information, but *better* information, i.e., more accurate, timely, and relevant for purposes of effective decision making, leading to better planning and control. This requires setting up a management information system based on careful investigation of its real purpose and contents.

3. *To improve efficiency:* The removal of bottlenecks, increased accuracy, reduced throughout time, elimination of unnecessary work, and a general toning up of the whole operation are the important factors here. As with the previous objectives, the true reason for each of the problems must be as-certained before attempting to impose a solution.

Examples of improvements in efficiency are the early detection of over-due debts, the speedier handling of paperwork for customers' orders, notifica-tion of excesses, obsolescence, or shortages in stock holdings.

Dictates of Top Management Policies

Policy with regard to a particular matter may be based on factors that are neither apparent nor disclosable to staff at a level lower than top manage-ment. Since this group includes the systems analyst, top management may be in the difficult position of trying to give a lead while at the same time being unable to disclose the full facts about a matter.

On other occasions the policy laid down by top management may appear to the systems analyst to be incorrect or not viable. He is entitled to his opinion, but he is not entitled to make surreptitious amendments to this policy when designing the system. It is not the function of either the systems analyst or the data processing system to make corrections to what appears to be erroneous policy. Attempts to do so may jeopardize the benefits of what is really a proper policy.

3.2 TOP MANAGEMENT SUPPORT AND INVOLVEMENT

For many firms the purchase of a computer is the biggest single financial transaction that its management has ever made. Although management might not be aware of it at the time, the decision to rent or to purchase a computer may well make a significant cost difference.

In spite of these two overwhelmingly important factors many managers have no appreciation of the computer's potential as a management tool. It has often been stated that lack of top management involvement is the most important contributory factor in the failure to use computers properly. In the early days of computers, management understandably adopted the attitude, "leave it to the experts." At that time the experts were the programmers, since they alone knew how to control the "electronic brain." What must be accepted nowadays is that it is the managers who should acquire the expertise on which depends the full value of data processing. The computer and the associated system are theirs, to be utilized to provide management information from the mass of routine data. This is not to suggest that management must learn programming; the necessary minimum is that they understand the computer's potential and the principles of data processing systems.

With these aims in view, a worthwhile first step is for top managers to attend a computer appreciation course. These are run by all computer manufacturers and by a number of consultant organizations and professional bodies. They usually involve full-time attendance at an education center for a few days. Although some of these courses may give the impression of attempting to indoctrinate, the experienced manager discounts this implication and perceives the real value of the subjects taught.

A second step for top management is to give full support to an investigation of the existing system and to make a close study of its findings. This also involves them in directing the course of an investigation so as to steer it toward their own aspirations. The results of the feasibility survey (Section 3.3) provide an early indication of the course being followed; by maintaining contact with the systems investigation and design, the course can be rechecked at regular intervals.

A third step for top management is to make a clear decision regarding adoption of the new system. This is based upon the systems analyst's report (Section 12.4).

A fourth step is to support, and to some extent become involved in, the implementation of the new system.

Systems Investigation Support

How can top management best give support at the investigation stage?

1. By providing the systems analyst with an assignment brief (Section 3.3) and being prepared to explain objectives, policy, and developments in and around the organization.
2. By officially informing all departmental managers of the impending feasibility survey and systems investigation, giving brief but convincing explanations of the purpose of these exercises. Also, where necessary, arranging for departmental managers to make available a few members of their staff to assist in the systems investigation as temporary members of project teams.

3. By helping to overcome any resistance to the investigation and, if necessary, using managerial authority to direct staff to cooperate.

4. By appointing one present member of the board (or its equivalent) to be responsible for management services, including systems analysis and data processing. The systems analyst, or the data processing manager when appointed, would report directly to this director, who would take a particular interest in the top management aspects of data processing.

The Steering Committee

The board member appointed to be responsible for data processing ideally should not have any bias toward a particular department. The new system will cut across departmental boundaries and involve integration of their work, at the same time requiring the utmost cooperation from their staff. In order to meet these requirements, it is advisable for a steering committee to be formed with the above-mentioned director as its chairman.

Other members of the committee might well be the following:

Management Services Manager (when appointed).
Data Processing Manager (when appointed).
Senior Systems Analyst (until appointment of either of the above).
Managers of the departments most affected.
Representative from outside organization such as computer manufacturer or consultant firm (if involved).

At certain times other persons can be temporarily appointed to the committee when the areas of work with which they are concerned are under review. As with all committees, it is advisable to restrict the membership to not more than about eight persons so as to facilitate decision making. The decisions made by the steering committee are those that affect the company from the data processing aspect. Decisions pertaining only to the internal working of the data processing department can be made without reference to the committee.

The terms of reference of steering committees vary slightly from one organization to another, but the main points are:

1. To establish a program of activities connected with the actual or possible introduction of data processing into the organization. This program includes the systems investigation, a study of its findings and the recommendations therefrom, the systems design, and the implementation of the data processing system.

2. To maintain a watching brief on the program activities, and to use committee influence in overcoming problems of an interdepartmental nature.
3. To advise on the possible purchase of a computer, to select firms to be invited to bid, and to act as the official negotiating body with prospective suppliers.
4. To appoint a data processing manager if one is not already designated, and to designate his duties and responsibilities.

3.3 THE ASSIGNMENT BRIEF AND FEASIBILITY SURVEY

An assignment brief and the consequent feasibility survey are inevitably tied together in that the one always has some effect upon the other. It is un-realistic to assume that a precise assignment brief can be drawn up to cover all possible points that may arise. The brief is usually created prior to the formation of the steering committee, and therefore it is the responsibility of top management to prepare it. The problem here is the improbability of management's appreciating the potential of data processing until the feasibil-ity survey has been carried out. Thus, there must be an interchange between the assignment brief and the feasibility survey, the former being amended perhaps several times as a result of the latter.

The Assignment Brief

The major features of an assignment brief are:

1. It is an authorization for the systems analyst to carry out an investiga-tion of existing systems. In this respect it is a "passport" into all relevant departments, and should be regarded as a request from top management, ask-ing staff to assist the analyst in his investigations.
2. Initially it states in general terms the objectives of the feasibility survey; then, after suitable amendments have been made, it leads to the detailed objectives of the full investigation.
3. It indicates any limitations to be imposed on the survey—these may refer to applications, areas of work, or locations. They should not be im-posed without good reason because they can severely restrict the potential benefits of a system. Any areas that are not specifically prohibited are as-sumed to be within the scope of the survey.
4. Reference is made in the assignment brief to any previous surveys that have been made in a similar field. It is unwise for top management to de-liberately conceal the results of previous surveys, even though they are out-

dated and were unacceptable at the time. Quite trivial suggestions contained therein might now be practical methods in rudimentary form.

5. It provides an indication of the top limit of capital expenditure or annual outflow that can be considered. This limit must be realistic in relation to the hoped-for savings in future years, and is one of the features most likely to be amended as a result of the feasibility survey.

The Feasibility Survey

This forms an interim stage between the assignment brief and the full systems investigation. In certain circumstances it is permissible to omit the feasibility survey and to proceed immediately with the full investigation. This is usually the policy when the area of the full investigation is limited, or when it is the intention to transfer a well-established procedure to data processing.

Generally, the feasibility survey is carried out by the systems analyst who will later be engaged in the full systems investigation and design. The time taken for the survey should be kept quite short, subject to achieving the required results. During the survey the main source of information is line (middle) management.

The main aims of the feasibility survey are:

1. *To determine whether the objectives stated in the assignment brief are reasonably attainable within the limitations and financial constraints imposed.* If not, what limitations must be removed or what constraints loosened?

An example of this sort of thing is the aim of creating a computer-based production control system without availability of the necessary basic data such as parts lists and machine operation times. These groups of data are likely to take a considerable time to accumulate, but until this has been done the system cannot become operational. There is, of course, no reason why the systems investigation should not proceed, provided the situation is recognized and the remedy is accepted and applied.

2. *To define the major problem areas so that the systems analyst can plan his strategy for the full investigation.* In a small company this means merely that he can draw up a rough time schedule for himself. In a large organization it could involve planning the best deployment of several systems project teams, taking into account the departments and applications involved, together with the caliber of the systems analysts participating.

3. *To find the areas where potential exists for making savings in money, time, or effort.* These areas are not necessarily the same as the problem areas, but their discovery and appraisement will also contribute significantly to the strategic planning of the investigation.

In certain organizations these potential savings areas are manifest, owing to the existence of some large intractable job such as the massive billing operations of gas and electricity companies. Without such an obvious application, the factors listed below could be given consideration when searching for savings:

Work involving large-scale repetitive clerical operations.

Areas where large amounts of information are filed, processed, and assessed.

Procedures that involve excessive amounts of manual or mechanical calculation.

Areas that lend themselves to the application of more scientific methods, thereby increasing efficiency by increased speed, greater accuracy, reduction of stockholdings and waste, and increased turnover.

4. *To approximate the time required for the full investigation and the cost thereof.* This again relates closely to the second aim in that the time taken will depend very much upon the number and caliber of the systems personnel engaged in the investigation.

5. *To discover whether specialist knowledge will be needed for the full investigation.* This requirement may be filled by experts enlisted from within the organization or from outside agencies such as consultant firms or computer manufacturers. Alternatively, the situation may call merely for an expansion of the systems analyst's knowledge or for additional training. Examples of these specializations are operations research techniques, statistical methods, mathematics, and investment appraisal techniques.

Feasibility Survey Report

In view of the restricted time allowed for the feasibility survey, grandiose reports are best avoided. It is satisfactory to report the findings in the form of an extended memorandum combined with detailed addenda covering any complex points. A detailed report is, of course, presented to top management after the full investigation has been completed (Section 12.4).

The memorandum should cover the aims of the survey and should conclude with recommendations for further action. Upon rare occasions, nothing is to be gained from further investigation if it is perfectly clear that the existing system cannot be improved upon. An appropriate recommendation in this situation is that another feasibility survey be carried out at a later date as a recheck of the system or if certain circumstances arise that indicate its inefficiency.

3.4 EXERCISES

Problem 1. Objectives

ATC, Inc., manufactures and markets typewriters and other small office machines. The company has been expanding erratically during the past ten years and now has a firm market base in the United States. Customer order volumes fluctuate from month to month, but the net result is an overall increase of sales volume from year to year.

Ideally, each of ATC's 15 depots holds a full range of products, with stock levels based upon rough estimates of future demands in the locality.

A customer's order is met, if possible, from the stock in the nearest depot, and where this is insufficient, one or more of the neighboring depots are asked by phone to fill the order.

ATC's three factories, although located fairly near each other, operate autonomously and manufacture different models of the various machines sold. Similarly, purchasing is done independently, even though the suppliers and materials purchased are often the same for the three factories. Production targets are based upon the previous year's accomplishments, since it is known that these are more or less constant. This results, however, in certain shortages and in overproduction of other items. Shortages lead to overly long production-cycle times and consequent delays in delivery to depots and customers.

Suggest what should be the long-term and short-term objectives of the ATC management. Short-term aims are considered first.

Solutions to Problem 1 (Short-term objectives). It is obviously unsatisfactory to attempt to meet customers' orders from locally held stock unless this is well controlled and recorded. Telephoning to various depots is time consuming, haphazard, and liable to error.

(A)   First Short-Term Objective: To create and maintain on a centralized basis an accurate stock record of finished products in each depot so that each customer's order can be handled automatically by the nearest depot with the product in stock, without any need to refer to the others.

Segregated purchasing of materials from the same suppliers by the three factories does not take advantage of bulk-buying prices and discounts, nor does it minimize freight charges.

(B)   Second Short-Term Objective: To introduce a centralized purchasing system so that gross material costs will be reduced to an absolute minimum.

The handling of customers' orders ties in closely with the centralization of depot stock records, and it is also necessary to cope with the fluctuating volumes of orders.

(C)   Third Short-Term Objective: To organize the order-handling routine on a flexible basis, linked to the depot stock records system so that less overtime work will be needed in peak periods, and less time will be lost in slack periods.

Solutions to Problem 1 (Long-term objectives). Orders are increasing year by year and also are fluctuating month by month. Unless this situation is assessed quantitatively, little can be done to meet this demand, either from stock or by increased production.

(D)   First Long-Term Objective: To build up and analyze sales statistics in order to create a basis for sales forecasting. This fund of data will enable the analyst to set more accurate production targets and estimate the higher rate of turnover of depot stocks so that realistic sales forecasts can be made.

Production targets should not be based merely upon what is attainable but upon what is both attainable and needed.

(E)   Second Long-Term Objective: To develop a production planning and control system linked to sales forecasting and production capacities. This would reduce shortages and overproduction by accurate computation of parts requirements, machine loadings, and assembly loads.

Problem 2. Steering Committee

Assume that the management of ATC Inc. decides to attempt to meet all the objectives stated in Problem 1 by introducing a computer-based system. Suggest the composition of a suitable steering committee, briefly giving the reasons for choosing each member.

Solution to Problem 2. The reasons for choosing particular members are:
(A)   Production director (chairman): Although the chairman should be un-biased, the biggest task by far is the implementation of a computer-based production control system, and this will need powerful management backing.
(B)   Production control manager: Same as item (A).
(C)   Sales manager: Sales analysis and forecasting is crucial.
(D)   Chief buyer: Centralized purchasing is essential to cost control.
(E)   Stock control manager: Centralized stock recording requires coordination with other departments.
(F)   Customer orders department manager: Reorganization of customer order handling must conform to new system requirements.
(G)   Senior systems analyst: A qualified systems person must supervise until an experienced data processing manager is appointed.
(H)   Member-at-large: This person may be a representative from any consultants organization that is involved or he may be a member of top management who is acting as liaison between production and the policy-making group.

CHAPTER **4**

SYSTEMS INVESTIGATION

When the outcome of the feasibility survey is known, top management is in a position to decide whether to authorize a full investigation of existing systems. If this is the decision, approval should be made official by publishing a memorandum or notice to this effect, after which the investigation can proceed.

At this stage the systems analyst is probably confronted by a large organization whose complexities may appear overwhelming. A full investigation can, if completely unrestricted, take an interminable amount of time and effort. The systems analyst, by using the information obtained during the feasibility survey combined with his own experience, will impose his own limitations on the depth of the investigation. This is necessary in order to obtain the required information within a reasonable period of time, without accumulating a large volume of irrelevant data. The investigation must not be too restricted, for an attempt to work to a tight time schedule would almost certainly fail because the analyst must depend on other people whose availability largely determines the rate of progress.

Where the organization under investigation is large, the systems investigation must obviously be carried out by several systems analysts or even a team of analysts. In any event, the work of investigating existing systems is greatly facilitated by the employment of staff experienced in the business of the company. When the systems analysts themselves do not have this advantage, it is best to assign experienced employees in each department on a temporary basis. By organizing project teams comprising systems analysts and these experienced employees, the investigation work can be split up into convenient projects. This arrangement means that close contact is maintained between the systems and user departments; it also engenders a feeling of involvement

34

among the prospective users, thus encouraging their acceptance of the new system. At this level of systems investigation, it is evident that the senior systems analyst will be mainly occupied in organizing and coordinating the work of the project team. Accordingly, he is unlikely to do much of the investigating himself.

4.1 FACT FINDING AND VERIFICATION

What Are Facts?

The definition of "fact" in the scientific sense brings in the meaning of truth and accuracy. *For the purposes of data processing, however, facts can be regarded as any information that is relevant to the existing or proposed system.* Our main concern is to be able to find all the facts, to separate them from opinions, and at the same time not to lose sight of interesting and potentially valuable ideas. Facts are the "bricks" from which the new system will be built—ideas are its "architecture." Provided facts and ideas are not confused in the mind of the analyst, both are of value in the design of the data processing system.

The systems analyst is also concerned with the accuracy of the facts given to him. In this context, accuracy can be evaluated in two ways:

Is a statement true or false?
To what degree is a figure or measurement applicable?

Some facts are never absolutely accurate, but are near enough for practical purposes. An instance of this is the number of employees working in a particular department.

Employees come and go, so there is no point in trying to obtain absolutely accurate information of this nature. If this were to be attempted, other facts must be introduced. For example: at 3 P.M. on August 20, 1973, there were 23 employees in the purchasing department, 21 of whom were actually at work. All that is of interest in a case such as this is that there are about 23 employees in the purchasing department, and—provided this figure is not likely to change considerably—it is satisfactory for the systems analyst's purposes.

There are, on the other hand, certain facts whose accuracy must be absolute, and others for which a maximum allowance can be made. The former relate to things that control the design of the data processing system; for instance, the structure of a code number may decide the technique subsequently adopted for accessing stored data. The "maximum allowance" facts relate to quantities, prices, rates, etc., that are to be processed by the computer. Because we are dealing with a dynamic organization and transient situations, these factors are nowhere near constant, and the best we can do is to make allowance for the maximum value that they may attain.

How Are Facts Gathered?

The facts and information required from the investigation are not always in a written or immediately usable form. It is sometimes necessary to create or deduce the required information from other ascertainable facts. Generally, this is a simple task involving nothing more than the addition of a few figures. To the existing system these deducible facts may be of no value and therefore have not been made readily available. For example, the total number of piece parts handled in a factory may not have been previously ascertained because the existing production control system deals with them in quite separate groups. However, in designing the new system, this information is needed in order to determine the amount of computer storage space required for piece-parts files.

Facts are obtained by four main methods: asking, observing, measuring, and reading. At the outset of the investigation the analyst must prepare the ground so that he is able to implement these methods in an efficient and logical way. In order to gain access to the appropriate people and places, he should start by approaching the departmental managers. Thereafter, as far as is possible, all contacts should be made via a more senior person, if necessary going to top management for help. In most cases this will not be necessary because sufficient contacts will have been established during the feasibility study.

Having gained access to a department, the analyst should break down the psychological barriers that are bound to exist between himself and the department's staff. The surest way of doing this is for him to establish amicable relations with members of the staff. With this in mind, the analyst is well advised to obtain a temporary working place near the department's center of activity. This not only creates rapport between himself and the staff, but also facilitates observation of the department in action.

Asking. Inquiry usually entails the verbal questioning of the appropriate members of staff, and careful notation of their replies. A reply might not be strictly pertinent, in which case a rephrasing of the question is called for. It is imperative that the systems analyst have sufficient knowledge of the topic under discussion to enable him to detect a divergence of understanding. An example of this might be a difference in the conception of raw materials as applied to the firm. The systems analyst may believe raw materials to be completely unformed materials, whereas the interviewee may assume that castings are also included. A misapprehension of this sort, if not detected early in the discussion, can lead to further confusion later. Part of the systems analyst's skill lies in his ability to ascertain precise meanings quite quickly.

As well as asking about the actual details of the present system, the analyst ought to obtain ideas and suggestions for improvements. These are then mentally sifted in order to separate mere wishful thinking from the seeds of genuine advancements. A little gentle probing often encourages people with valuable ideas to make them known. The feeling that changes are in the offing, and the presence of an attentive listener, bring forth suggestions that might otherwise never be heard. People who seem to be completely devoid of ideas, if given the stimulus and some time to think, can sometimes produce previously suppressed but valuable ideas.

Questionnaires. An alternative means of asking is the questionnaire. This method is suitable for situations in which a large number of people are to be asked a number of straightforward questions. These would call for brief answers that can be recorded by the recipient of the questionnaire in a form that precludes subsequent misinterpretation. The questionnaire must be carefully designed so that the nature of the required information is clearly indicated. Great care is necessary in the phrasing of the questions, and they should be accompanied by a brief explanation of their general purpose to orientate the reader. This method is unsuitable for eliciting facts related to an actual system; the questions are too limited, and the replies often not forthcoming for a considerable time. A possible application of the questionnaire within systems investigation is as a means of interrogating widely deployed staff. These people might, for example, be employed in the company's branches or depots in various parts of the country or the world. In this case, a questionnaire can be used if the questions are suitable and the required information is not available from a single source.

Observing. The less of an observer and the more of a participant that a systems analyst can become, the more realistic will be the picture he obtains. His acceptance as a temporary member of a department will enable him to observe while being part of the system. This facilitates a study of the actual situation, the department in action, and the methods for dealing with exceptional conditions.

Actual observation also gives the systems analyst the opportunity to assess the following characteristics of a department:

1. The pressure of work: high or low, steady or variable, isolated or evenly spread among the staff.
2. Movements of personnel: within the department and between it and other departments. This movement may be a means of transmitting information between departments (Section 4.3).
3. The attitude of the staff toward the existing system: favorable, unfavorable, resigned acceptance, or possibly not really understanding it.
4. The volume of telephone calls: callers from other departments, visitors from outside the company, or other interruptions to routine.

5. Usage of files: for routine purposes, answering queries, or other special reasons.

Measuring. The main purpose of this activity is to approximate the amounts of documents, items, persons, transactions, etc., concerned with each sphere of work. This information is normally obtainable from the appropriate staff, but when it is unobtainable in this way, or is suspect, actual measurements or estimations must be made. If this is done by sampling, care should be taken to ensure that truly representative samples are used.

Another purpose of measuring is to ascertain the times taken for certain activities and the frequency of their occurrence. These times include not only the actual task-performance times but also the intervals between the completion of a task and its effect elsewhere. It is not recommended that all tasks and intervals be carefully timed; the need is for the approximate times related to the main activities.

The frequency of activities applies in two respects. First, the frequency with which a routine is carried out—for example: invoicing is daily, payroll is weekly. Second, the rate of occurrence of variable events—for example: customers' orders average 250 per day, January through August, rising to a peak of 500 in November before falling back by January.

Reading. Much of this topic is covered in Section 4.2 (Inspection of Records), but reading also includes other information such as:

1. Reports of previous surveys and investigations: These are worth studying carefully, but any facts therein that appear to be usable should be confirmed by other methods. It is not satisfactory merely to inquire about the general validity of a report; its originators are almost certain to confirm this, regardless of changed circumstances. Recent reports are the most helpful, but outdated ones may also provide useful ideas.

2. Company instructions, memoranda, and letters: These are likely to be voluminous. It is therefore wise to persuade others to extract the appropriate documents from their files for the systems analyst's perusal.

3. Company information booklets and sales literature: These are worth scanning in order to obtain general information about the company's activities and connections.

Verification of Facts

Ideally, the systems analyst obtains verification of every fact that he collects. This is not always possible in practice, but nevertheless he must attempt to do so for all facts obtained verbally. The answer to a question can be erroneous for one of three reasons: ignorance of the subject, leading to a mistaken answer; a misunderstanding of the question, provoking an irrelevant answer (sometimes not discernible as such); and a deliberate misrepresentation of the situation.

The third of these reasons is fairly uncommon and is not usually associated with quantitative facts.

How can these difficulties be dealt with? The obvious method is to ask another person with access to the same information and then compare the two answers. If the answers seriously conflict, their reconciliation requires a little diplomacy on the part of the analyst in order to avoid exhibiting someone's ignorance or mistake. Quoting the first person's answer in order to obtain its confirmation is not altogether wise, as this tends to encourage vague replies. A better approach is to restate the question, more clearly if necessary, to both persons and to look for substantiation within the answers.

Another method of verification is to ask a more searching question of the same person after he has given his answer to the first question. For instance, having obtained a suspect answer regarding the number of raw materials handled, the second question might be, "In what way are raw materials subdivided, and how many are there in each group?" The answer to this not only provides valuable information in itself, but also confirms the meaning of raw materials and verifies the accuracy of the number previously quoted. Typically, the first answer might mention the existence of 200 different raw materials, and the second answer might be that they are subdivided into 30 sheet metals, 100 bar metals, and 70 rough castings. Thus, it would then be immediately apparent as to what was really included and the total of 200 would be substantiated to some extent. On the other hand, the interviewee might realize that his categorized figures do not add up to the original total stated, and hence additional investigation or explanation would be required.

When a large amount of important information has been obtained verbally, written confirmation should be requested. If this is done informally and with an explanation of the reason, the recipient of the request is not made to feel in any way disturbed.

In addition to the verbal means of verification, the analyst may sometimes employ one or more of the other methods mentioned previously, i.e., observing, measuring, reading. As a general principle, two of the four methods should be combined in order to obtain and check any set of facts.

What Facts Are Needed?

If the ranges of organizations and applications suited to data processing were restricted, a compendium of suitable questions could be usefully compiled. Although this compilation is not possible in a comprehensive way, it has in fact been done in a limited form by several computer manufacturers for the benefit of their salesmen. These "checklists" cover the more common applications, such as payroll, sales accounting, sales analysis, and stores control. The questions on them are necessarily stereotyped, but nevertheless they are convenient reminders, and a novice to systems investigation would find them useful.

The required facts, in addition to those that are described in subsequent sections, can be broadly categorized as listed below.

1. *Ranges of Existent Items.* In this context, ranges of items include products sold, components manufactured, persons employed, suppliers dealt with, machines used, and so on. The list of ranges is likely to be quite extensive and to cut across interdepartmental boundaries. The information needed includes

How many items are there in each range?
In what way and for what reasons are they subdivided within each range?
Is a range subject to sudden changes of size or content, and if so, why?
What other changes apply to each range? Are these gradual, permanent, or such as to make a fundamental alteration to the range of usage?

EXAMPLE A. The range of suppliers consists of 60 companies subdivided into 35 local firms, 20 distant United States firms, and 5 foreign firms. This range is stable, with only one or two changes each year.

EXAMPLE B. The range of commodities sold at present numbers 700, of which 500 are for home consumption, 50 for export only, and 150 for both home and export. This range changes suddenly twice a year in content, size, and subdivision, and is tending to increase steadily from year to year.

2. *Information about Items*
What descriptions, quantities, and values are associated with each type of item?
Are the descriptions standardized?
What are the maxima and minima of the quantities and values?
Which of the quantities and values are liable to fluctuate, and what causes this?

In this context, "descriptions" include names and addresses as well as the conventional meanings. "Quantities" cover stock levels, order quantities, sales-tax percentages, discount percentages, reorder levels, and any other nondollar figures. "Values" include all amounts expressed in dollars, such as prices, wages, income tax, and costs.

3. *Code Numbers Used.* A range of items may have a set of code numbers

associated with it. Where this is so, it is important to check that this set is complete, consistent, and nonduplicated. The information of interest here is the precise layout of each code number, its limits as to content, and the meaning (if any) of the various digits, letters, and symbols therein. For example, each of a range of 1500 company customers has been allocated an account number within the spread from 10,000 through 99,999. The first two digits signify the location of the customer's head office (10 through 61 represent states; 65 through 83 represent principal foreign countries; and 90 through 99 represent other foreign groupings). The maximum number of customers within any one such subdivision is 120, although it allows for up to 1000. The final three digits of the account numbers are merely allocated serially to customers within each subdivision, and have no intrinsic meaning.

Other questions of pertinence are:

To what extent is the code number used externally to the organization (for example, do the customers themselves use these account numbers? What would be the effect of changing or redesigning the code numbers?

In general terms, it is wiser not to alter large sets of code numbers. There are occasions, however, when this is advantageous: For example, when an uncoded range is encountered, it may be necessary to introduce a new set of codes. Then a careful study of the characteristics of the items in the range might yield subsequent benefits (see Chapter 5).

4. *Calculations Performed.* These are not complicated in business applications, but can be regarded as including all operations involving quantities and values. For example:

What fractions and decimals are in use in connection with weights, lengths, areas, volumes, quantities, and money?
Do the calculations involve any special problems connected with foreign currencies or unusual weights and measures?
When extending (multiplying), how is rounding-off performed, and does this involve any reconciliation problems?

5. *Amounts of Data Handled*
What amounts of data are received from and dispatched to outside organizations and internal departments?
Are these amounts reasonably constant, or do they fluctuate significantly?
If so, to what extent and in what way? Fluctuations may be seasonal, monthly, weekly, or on some other regular cycle, or at random.
Are the amounts of data increasing? If so, at what rate, and is this likely to continue?

Where it is the intention to operate the new system on a real-time basis, the timings associated with data play a more prominent role because of the need for immediate response. Additional questions have to be asked and the replies verified carefully, actually measuring the relevant data if at all possible.

What are the arrival patterns of each type of input data, i.e., transactions or messages?

When do peak periods occur and for what reasons, and what levels of message rates are then reached?

How long is each type of input message, and what amount of output information or length of reply message will it call for?

How quickly does the user require replies to his messages, i.e., what response times are needed?

The design of real-time systems incorporating interactive (conversational mode) procedures needs even more information regarding the exchange of messages between the terminal user and the computer. The systems analyst needs to find out the structures of the separate messages that will occur during the man/computer conversation, and the frequency with which these messages will be transmitted each way. It is unlikely that this information will be forthcoming merely as a result of asking for it. The prospective user of conversational mode needs to have the general procedure carefully explained to him and then must devote considerable time, in conjunction with the systems analyst, thinking about the precise procedure he wants.

Irrelevant Data and Information

Desire on the part of the systems analyst to make the investigation thorough and comprehensive engenders the collection of large amounts of data. Some of this is bound to be irrelevant, and an awareness of this possibility enables the analyst to avoid irrelevancies, thereby not only saving time and effort but also avoiding the obscurement of significant facts upon which the new system will be based. During the investigation the analyst is interested in information about the data rather than in the data itself. A study of a firm's products, for instance, should yield quantitative and qualitative information of their classification, manufacturing control, sales turnover, and so on, rather than merely a list of the products.

Information regarding methods and policies is not easily categorized as "important" and "unimportant"—what is unimportant today may be extremely important tomorrow. A piece of information that is insignificant in one situation can be of vital interest in another. For example, the impending retirement of the cost accountant may well become a valuable piece of information when considered in relation to plans for a new costing system.

With the gaining of experience, the analyst can avoid the accumulation of trivia without, at the same time, losing any essential information.

4.2 INSPECTION OF RECORDS

The term "record" as used in this section refers to anything on which facts and figures have been written or printed. It includes ledgers, lists, catalogs, interpreted punched cards, forms, and other documents of a similar nature.

Meaning of Entries on Records

During the course of an investigation the systems analyst acquires copies of a wide variety of records in the form of documents of various sizes, shapes, and colors, which contain a multitude of different entries. It is therefore imperative that all these entries are discerned and understood not only at the time of acquisition but also later when the new system is being designed. The majority of records start life as blank sheets containing only their preprinted headings. With the passage of time, the changes inherent to any organization cause some of the headings and their associated entries to be at variance. To a comparative stranger to the system these variances may not be at all obvious. The systems analyst must therefore be watchful for:

Entries under incorrect headings.
Entries never made, despite the headings.
Entries made for which no heading exists.

Detection of these irregularities is facilitated by comparing "live" records holding real data. These can be inspected at leisure if photocopies are made of current records, or if specimens of recently used records are obtained. By inspecting live documents, the systems analyst can judge their legibility, bearing in mind their possible future use as source documents. He should, however, carefully check the vintage of supposedly live documents in case they are so old as to be misleading.

Record Specification Form

When inspecting records, the systems analyst should find the meaning of any entry that is not obvious, making a note of any special symbols, entries in colored writing, and remarks. For all but the simplest records, it is advisable for him to fill in a "Record Specification Form" as a means of accurately describing the record's entry data. An example of this form is shown in Fig. 4.2; this has been filled in for the record shown in Fig. 4.1. Most of the entries on the record specification form are self-explanatory, but a few need a little more explanation.

Reference Number. This is any reference on the record that identifies it; if there is none printed, a number should be entered by the analyst on his specimen so as to connect it with its record specification form.

Entry Reference. This is a reference written and circled on the specimen record by the analyst so as to identify each entry thereon.

Picture. The figures and letters in this column indicate the size and layout of the record's entry and provide information for use later when designing a computer-based system. The meanings of the characters in the picture are as follows:

Character	Meaning
9	Numeric digit between 0 and 9
3	Numeric digit between 0 and 3 and so on
½	0 or ½
¾	0, ¼, ½, or ¾
A	Alphabetic letter from A to Z
B	Blank
.	Decimal point
X	Any of the above and also symbols
9(5)	Any 99999
X(4)	XXXX, XXX, XX, or X

Examples of pictures

X(15)	A field of up to 15 alphabetic letters, numeric characters, blanks or symbols; for example: JOHN HARGREAVES, or ACACIA AVENUE, or SHOE/BROWN 8½
999.9	A field of three whole numbers and one decimal; for example: 125.7
A(6)	A field of up to 6 alphabetic letters; for example, WASHER or BOLT
AAAB99A	Three alphabetic letters, a space, two numeric digits, and finally one alphabetic; for example: FWD 36C

Sometimes several pictures are applicable to one entry; for example, code numbers deriving from separate external sources such as state car-registration numbers. In this case, each picture should be shown on the record specification form with a brief explanation under "remarks."

Maximum. This is simply the highest value that an entry can attain. In many cases it turns out in practice to be the same as the picture, but is sometimes less than this; for example: entry reference K (no. of invoices has a pic-

ORDER FROM:
Pelham Caravans Inc.
Pelham Street,
Newark, N.J.   (A)

INVOICE TO:
Same

DELIVERY ADDRESS:
Pelham Caravans
Mills Drive,
Farnden Road.
Newark, N.J.   (B)   (C)

CUST. ORDER No.: Verbal   (D)
TELEPHONE No.: 754-3286   (E)
CONTACT: Mr. Pollard   (F)

No. 3590   (G)
SALESMAN'S SIGNATURE: J.W.   (H)
DATE 9/8/73   (I)

ACCOUNT: Monthly   (J)   No. OF INVOICES 1   (K)   DELIVERY METHOD O/V   (L)

Item (P)	Quantity (Q)	Description (M)	Our Price (R)	Weight (S)	Supplier (T)	Our Order No. (U)	Supplier's Price (V)	Supplier Delivery (X)	Our Invoice No. (Y)	Invoice Value (Z)
		Aluminum Alloy HE 30WP								
	2	23' SR12 Side Rave (Flat)	.58	2 x 23 ft / 59 Lbs (N→59)	Impalco	682	.57	x/s		34.22
	2	24' 4" x 2" Lipped channels (83)	.57	2 x 24 ft / 81 Lbs	Impalco		.56 -5% (W)	x/s		46.17
	2	24' 9" x $\frac{1}{2}$" corrugated planks (51)	.65	2 x 24 ft / 50 Lbs	Impalco		.64 -5%	x/s		32.50
	3	15' 3" x 1$\frac{1}{2}$" x $\frac{3}{16}$" channel (164)	.52	3 x 15 ft / 83 Lbs	Impalco		.47 -5%	x/s	1359	43.16
	3	16' 3" x 1$\frac{1}{4}$" x $\frac{7}{16}$" channel		3 x 16 ft / 87 Lbs	Impalco		-5%	x/s		45.24
	4	16' 1" x 1$\frac{1}{4}$" x $\frac{1}{8}$" angle (20)	.62	4 x 16 ft / 22 Lbs	Impalco		.61 -5%	x/s		13.64
									(AA →)	214.93

Fig. 4.1 Specimen documentary record

RECORD SPECIFICATION FORM

NAME OF RECORD			DEPARTMENT	
Order sheet			Sales	

| REFERENCE No. PFC 9171 | DATE FILLED IN 2/15/70 | | FILLED IN BY HDC | |

ENTRY REF.	HEADING	PICTURE	MAXIMUM	ENTERED BY	REMARKS
A	Order from	30 char/line	100 chars	Salesman	Up to 5 lines
B	Invoice to	30 char/line	100 chars	Salesman	Usually same as 'A'
C	Delivery address	30 char/line	100 chars	Salesman	Usually different from 'A'
D	Cust. order No.	10 chars	10 chars	Salesman	Usually blank (verbal)
E	Telephone No.	AAA 99999	—	Salesman	
F	Contact	20 chars	20 chars	Salesman	
G	No.	99999	99999	Preprinted	Sequential ref. No.
H	Salesman's Sign'r.			Salesman	
I	Date	12/31/99		Salesman	Date order received
J	Account	7 chars		Salesman	Monthly or cash
K	No. of invoices	9	5	Salesman	= No. of order sheets
L	Delivery method	4 chars		Salesman	Limited range
					(O/V, BR, BRS, Coll)
M	Description	30 chars	30 chars	Salesman	Isolated description
					refers to items following
N	Length (per piece)	99	30 ft	Salesman	Circled figure = est. weight
P	Item	B			Not used
Q	Quantity	999	500	Salesman	Pieces ordered
R	Our price	2.40	1.80	Salesman	Per lb.
S	Weight	999	200	Cost clerk	From which inv. val. calc.
T	Supplier	10 chars		Purch'g clerk	Limited (could be coded)
U	Our order No.	9999	9999	Purch'g clerk	
V	Supplier's price	2.40	1.50	Cost clerk	Per lb. ± discount
W	(Discount)	± 9½	± 7½	Cost clerk	+ if surcharge
X	Supplier delivery	X/S or 99	15	Purch'g clerk	X/S = immediate, or weeks
Y	Our invoice No.	9999	9999	Cost clerk	
Z	Invoice value	999.99	$960	Cost clerk	= item value
AA	(Total value)	9999.99	$480	Cost clerk	On last sheet only

Fig. 4.2 Record specification form

ture of "9" but a maximum value of only 5; other similar examples are entry references N, Q, R, S, V, W, X, Z, and AA.

4.3 INFORMATION FLOW

The transmission of information within an organization is a means by which control and stability are achieved. It is therefore vital that no existing channels of communication be overlooked during the investigation of the present system. Since the main method of transmitting information and data is by means of documents, a careful study of document movement is essential.

Document Movement

Documents originate, both within and outside the organization, at the time when the first entries are made on them. Thereafter, further data is recorded and extracted at several points before the records are finally filed away, destroyed, or dispatched to outside organizations. By tracing these movements and recordings, the systems analyst can build up a picture of the information flow within the organization. During his visits to the various departments, the following questions must be answered:

> What documents originate here and how many copies are prepared?
> To where are these documents dispatched?
> What documents are received from elsewhere?
> What documents are filed and for how long are they filed before being disposed of?
> What entries are made on each document and what work does this involve?
> What information is extracted from each document and for what purpose?

By cross-checking the answers received from the various departments, omissions and misunderstandings are identified.

Document Movement Form. This form is complementary to the record specification form (Section 4.2), but is applicable only to records that move between departments in the form of documents. The document movement form is less susceptible to standardization than is the record specification form, and is best designed to meet the precise needs of each individual movement pattern. It would, however, contain all or some of the following information:

1. *Identification:* The document's name, title, and/or reference number (tying in with the corresponding record specification form).

2. *Purpose:* Concise details of its main uses.
3. *Origin:* Where and by whom it is originated, the number of copies made, and the differences (if any) between them.
4. *Distribution:* To whom each copy of the document is distributed after origination, and brief details of its movement thereafter, referring to the relevant flow charts (see below).
5. *Volumes:* The average and peak rates at which the document is created, and the average number of entries on it if these are variable; the number of documents in existence in each department at any point of time; a note of any cyclic or seasonal variations in the above figures.
6. *Sequences:* Any significant sequences in which the documents are moved or are filed in each department; these may arise naturally or through sorting.
7. *Special considerations:* Any further points of importance not included above.

Flowcharts of Existing Routines

During the systems investigation the analyst pieces together the pattern of operations within each department's routines, and relates this to the movement of documents between departments. This pattern can be represented in a number of ways, but for most cases it is best done in diagrammatic form. This entails preparing a "flowchart" such as that shown in Fig. 4.3. This example relates to the order-handling routine of a firm of aluminum stocklists, and ties in with the "order sheet" described in Section 4.2 and Figs. 4.1 and 4.2.

On a flowchart each operation within an existing routine is represented by a box that is numbered arbitrarily for reference purposes. Within each box is a brief explanation of the operation, together with the names and/or reference numbers of any records that are involved. If the operation is too complex to describe adequately within the confines of a box, this can be done on a separate sheet and referred to in the box. Arrows indicate the general sequence of the operations and allow for cross-referencing between flowcharts. The precise layout of the flow chart is not as important as its clarity, and provided all the systems analysts in the one organization follow the same conventions for drawing and numbering, no confusion will arise. Each flowchart has a reference number, and it is often convenient to code this so as to indicate the department (or application) to which it applies. The flowchart in Fig. 4.3, for instance, has the reference number S4, the S indicating that it applies to a sales department routine.

When preparing these charts, it is advisable to allow plenty of space on them and not to put too many operations on one sheet. This allows for further insertions or remarks that may be necessary later.

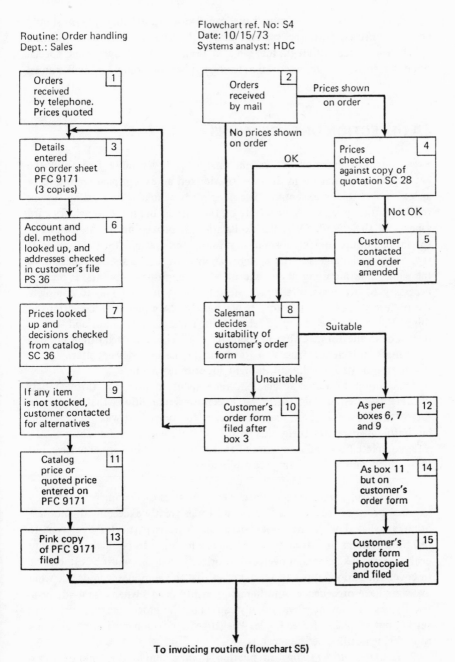

Routine: Order handling
Dept.: Sales

Flowchart ref. No: S4
Date: 10/15/73
Systems analyst: HDC

1 Orders received by telephone. Prices quoted	**2** Orders received by mail

Prices shown on order

No prices shown on order

4 Prices checked against copy of quotation SC 28

OK

3 Details entered on order sheet PFC 9171 (3 copies)

Not OK

6 Account and del. method looked up, and addresses checked in customer's file PS 36

5 Customer contacted and order amended

7 Prices looked up and decisions checked from catalog SC 36

8 Salesman decides suitability of customer's order form

Suitable

Unsuitable

9 If any item is not stocked, customer contacted for alternatives

10 Customer's order form filed after box 3

12 As per boxes 6, 7 and 9

11 Catalog price or quoted price entered on PFC 9171

14 As box 11 but on customer's order form

13 Pink copy of PFC 9171 filed

15 Customer's order form photocopied and filed

To invoicing routine (flowchart S5)

Fig. 4.3 Flowchart of existing routine

The flowcharts prepared during the investigation of existing systems should not be confused with the systems design flowcharts drawn later (Section 10.5). The latter, although the same in principle, are drawn using special symbolic shapes to represent the various types of computer and manual operations.

4.4 DETECTION OF EXCEPTIONS

Even the most experienced systems analyst is never quite free from the nagging anxiety that he might have overlooked an exceptional condition or an exception to the exception. The people who explain the existing system to him naturally tend to concentrate on the normal and more straightforward aspects of their work. A verbal explanation, if broken up by discussions of special conditions, quickly loses its continuity and meaning for the listener. It is therefore advisable for the systems analyst to restate the explanation, at the same time inquiring about exceptions at appropriate points. It may be necessary to do this several times, steadily filling in the details to complete the picture. It is not satisfactory to ask merely, "Are there any exceptions to this?" Rather it is necessary to select particular points of doubt and to pose questions in the form, "What happens if . . . ?" This approach brings unusual conditions and circumstances to the mind of the interviewee; alternatively, it may trigger off a new line of thought about the system.

In addition to verbal discussion, time spent in browsing through past documents often highlights former exceptional conditions that have been forgotten. These, however, may have been temporary and will not reappear, but before disregarding them, this must be established beyond doubt.

A complete list of all categories of exceptions would be very extensive, but those most commonly existing are outlined below:

1. *Abnormal loads and amounts:* Peak amounts of documents caused by seasonal or other conditions; figures and totals greatly exceeding the normal, often associated with the above situation. This is particularly relevant to real-time systems in that complex arrangements have to be made in order to deal with various degrees of overload. A priority structure must be established to cover the various types of messages so that the computer can decide what messages have precedence. Another abnormality occurs when essentially positive figures become negative, usually caused by the processing of transactions getting out of phase; for example: "negative" stock-in-hand caused by stock issues being notified before receipts.

2. *Special considerations:* In relation to new employees, unknown cus-

tomers, introduction of new lines, etc., which need the implementation of special tasks before they are absorbed into the system; priorities given to certain customers or jobs (these priorities may lose their point with the introduction of a data processing system); allowances made outside those formulated (for example, bonuses, prices, and discounts decided manually at the time and place of application).

3. *Periodic additional work:* Extra work just prior to statutory holidays, especially for payroll staff in preparing several weeks' pay at the one time; stocktaking and stock evaluation carried out at regular and irregular intervals; financial year-end work and other compulsory information; periodic and ad hoc reports for management.

4. *Missing data:* Entries omitted unintentionally from documents, especially from those originating outside the firm. What are the arrangements for ascertaining and entering these omissions? When the data from these source documents is input to the computer in on-line mode, the missing data is immediately detected by the computer. Nevertheless, the exact details of both the obligatory and optional data must be determined during the investigation so that this detection procedure can be built into the computer's program.

5. *Mistakes:* Nonagreement of control totals (what procedures exist for their reconciliation?); arithmetic and transcription errors (how does their correction affect balancing and auditing?); code numbers (are methods available for checking these? See "check digits" Section 5.5). As with missing data, an on-line real-time system is able to detect certain mistakes in the input messages and reject or query these immediately. It is therefore important to ascertain what mistakes are likely to occur and how these can be best corrected.

4.5 HUMAN ASPECTS OF THE INVESTIGATION

A mention of the word "investigation" immediately engenders in some members of staff a feeling of apprehension. If followed by terms such as "computer systems" or "data processing," this apprehension rapidly develops into a state of acute anxiety. These conditions are really caused by two main factors: first, a fear of the unknown; second, fear of the possible loss of status. The latter culminates in a genuine fear of being made redundant; this is especially prevalent among older clerical staff. The fear is, more often than not, without foundation, since there is usually an overall shortage of clerical staff within the organization. This means that they can be absorbed by other departments if made redundant by the data processing system, such as employment as an operator, programmer, or control clerk within the data processing department itself.

The systems analyst can start to allay these fears during his interviews with the members of staff in the investigation stage. The "unknown" factor may take several forms, but is generally the computer, the new system, new types of staff, or the company's future policy. A brief discussion of these points will put them in proper perspective, and although the analyst cannot be expected to provide all the answers at this stage, the clerical staff will feel happier in the knowledge that their interests are being given consideration.

A general explanation of the purpose of the investigation, as issued by management and reinforced by the systems analyst, is invaluable. A discussion of the impending changes gives the analyst an opportunity to establish rapport with the departmental staff, and to encourage them in their desire to be involved in the new system. To suggest to staff that involvement with data processing will very probably increase their status and potential is in no way deceitful. The degree of interest shown by each member of staff is usefully noted for later consideration during the recruitment of the data processing staff.

At the investigation stage, it is often impossible for the analyst to give accurate replies to direct questions about a particular person's future prospects. It is better for him to be noncommittal than to administer "soothing syrup" in the form of vague blanket assurances. These are apt to backfire on him when radical changes occur later and at the time when the utmost cooperation is needed from the staff.

In situations where it is definitely not the intention to attempt to reduce staff numbers by introducing a computer, it is helpful if the company's management issues a statement declaring that there will be no layoffs or transfers. Obviously, this happy state of affairs cannot always be retained, and so an honest declaration of projected staff changes should be made. This will enable redundant employees to find new jobs in good time; those who are requested to stay on until the computer becomes operational should be adequately compensated.

Those members of the user department staff who will be working in cooperation with the data processing department must, of course, be adequately briefed (see Section 13.4), and the earlier in the chain of events that they are made to feel involved, the better. To paraphrase a well-known saying, "Nothing interests like involvement." Even during the early stages of systems investigation, interest in data processing can be stimulated by showing films, giving brief talks on the subject, and arranging visits to computer departments in other firms. The computer manufacturers are able and willing to assist in these programs.

4.6 INTERVIEWING

Interviewing forms an essential part of a system investigation, and should be thought of as discussions rather than as formal interrogation sessions. An interview that acquires the semblance of a formal eye-to-eye confrontation is doomed to failure. The systems analyst must always bear in mind that he is the potential beneficiary from the interview more than is the interviewee.

It is necessary during an investigation to interview the appropriate members of staff at all levels. The analyst's approach must therefore be flexible enough to deal with all types of personalities and situations.

All interviews should be arranged in advance to take place at a time when there will be a minimum of interruptions. If during the course of an interview there is a continuous stream of interruptions, it is wiser to conclude it prematurely than to allow it to disintegrate; it can then be reinstated later under more favorable conditions. This also applies when the interviewee is under obvious stress, caused by either the interruptions or another factor, such as a sudden and unexpected rush of work.

The levels of the staff with whom contact is made during an investigation vary from one organization to another. A person's title conveys little meaning in itself, and can be assessed only in relation to the size and structure of the organization in which he works. Broadly, we can consider four levels of staff, and the most suitable approach is adopted accordingly. These levels, from the systems aspect, are:

Top management (who decide policy).
Middle management (who control procedures).
Skilled staff (usually senior clerical staff).
Other staff (who follow a routine pattern of work).

Top Management Interviewing

Contact between the systems analyst and top management will have already been made during the assignment and feasibility stages. Following upon this initial contact, every effort must be made to obtain individual as well as collective views. What information should be looked for from this level? The analyst cannot merely adopt the role of questioner and thus obtain all he requires. He is as much concerned with hearing opinions as in obtaining straightforward answers. He should not be overawed by top management, remembering that they are likely to be just as much in awe of new systems involving computers.

The main points to be covered in interviews with top management are as follows:

1. Take every opportunity to cultivate the seeds of top management's comprehension of data processing that were sown at the feasibility survey stage.
2. Obtain ideas, opinions, and facts in relation to objectives, competition, and major problems, and explain their relevance to the new system.
3. Avoid asking for trivial facts or precise details of methods.
4. Ascertain the future strategy of the firm as regards mergers, takeovers, marketing, and manufacturing policies. Assurances should be given by the analyst that this information will be treated as confidential, but nevertheless it may not always be forthcoming.
5. Inquire about the structure and administration of the company, particularly as regards the relative positions and responsibilities of line management.
6. The employment of a tape recorder during the interview might be acceptable at this level; permission should be obtained first, however.

Middle Management Interviewing

The points to be included are as follows:

1. When making arrangements for the interview, give the manager some idea of its subject and purpose.
2. Before the interview takes place, brief oneself on the general duties and position of the interviewee, and also on the topics to be discussed. Vagueness on the part of the analyst results in the manager's loss of respect for him.
3. Try to involve only one other person at a time. If the members of the manager's staff are called in to answer specific queries, get them to depart as soon as this has been done; alternatively, postpone their intervention until the end of the interview. Do not ask the manager's staff, in his presence, questions that have already been put to him, even if he failed to provide an answer. These are better asked later when interviewing the staff privately.
4. Avoid questions concerning higher-level policy, but discuss that which he formulates. Ask for suggested improvements to the existing system.
5. If the manager is constantly interrupted, suggest that the interview be held elsewhere than in his office; for instance, in the analyst's office or in another quiet and private room.
6. Ask about the duties and responsibilities of each senior member of his staff, and make discreet inquiries regarding their personalities so that any difficult individuals can be anticipated. Seek permission to interview his staff, explaining briefly the reason for this requirement.

7. Control the interview: avoid wandering too far from the subject; do not allow generalizations to obscure the actual situation; separate opinions from facts.
8. Conclude the interview with a quick résumé of the ground covered, leaving the door open for further contact, and confirm doubtful facts by later sending a memorandum to the interviewee.

Skilled Staff Interviewing

With these people, the points to bear in mind are as follows:

1. Restrict the questions to details of the individual's duties only; avoid discussing policy and the posing of leading questions.
2. Show a competent interest in his work, avoid condescension, and do not make adverse comments.
3. Do not allow his manager or other members of staff to be present all the the time; discussion groups are more useful at a later date.
4. Take notes, and use a tape recorder, explaining the reasons for these to the interviewee.
5. Allow time for him to collate any required lists and copies of documents, returning a few days later to collect these if necessary. It is better not to allow the interview to become broken up by searches for documents but instead to collect these afterwards.
6. Do not attempt to cover too much ground in one interview; one hour at a time is sufficient. If necessary, make an appointment to continue the interview at a later date, preferably within a few days. Other points are the same as 4, 7, and 8 of middle management interviewing.

Other Staff Interviewing

It is not usually necessary to interview more than a small proportion of the other staff, as it is almost inevitable that the details of their work will have been already explained by the more senior staff.

The points to remember are, however:

1. Keep the questions simple, and do not encourage the stating of opinions.
2. Allow him to demonstrate his work rather than explain it verbally, obtaining copies of the documents he uses.
3. Be friendly, adopting a neutral attitude toward his relationship with management if this subject arises.
4. Show interest in his work, avoiding condescension and adverse comments, and be complimentary whenever possible.
5. Their efforts may be valuable during the implementation stage.

4.7 STAFF ORGANIZATION AND UTILIZATION

Staff Organization

The overall structure of the organization will, by this stage, be well known to the systems analyst. With this knowledge the relevant departments can be chosen for further examination and their structure put into the form of a departmental organizational chart similar to the example shown in Fig. 4.4. These charts not only help in the planning of the investigation, but are also used to prepare a staff establishment table, as described below.

A departmental organizational chart has a box for each manager and supervisor, and each is connected by lines to that of his superior and to the sections that he controls. A manager's or supervisor's box contains his title, name, and possibly other information, such as his location. Each section box contains the name or duties of the section, together with the number of staff in each "grade." The precise meaning of grade as used on this chart cannot be universally defined, but should be decided in relation to the size and type of company or organization. In some large companies, and in local and national government departments, staff grades are already defined; otherwise the staff can be "graded" unofficially by the analyst. Three or four grades are sufficient and their allocation must be confidential, since this is not in any way an attempt to compare staff individually, but is merely a means of obtaining an idea of the structural balances of departments. It must be appreciated that in some organizations the preparation of this chart will not be a straightforward matter, particularly if the staff is organized nonhierarchically.

The departmental structure charts are summarized and combined to form a staff establishment table as shown by the example in Fig. 4.5. This table is then used to provide a picture of the spread of staff, and to check against the payroll distribution that all personnel have been accounted for. It also indicates the areas for potential staff economies and improvements in efficiency.

Staff Utilization

When questioned on the subject, very few persons will admit that they are underoccupied, even though it is well known that time is being wasted in the department as a whole. Nevertheless, a rough measure of staff utilization can be obtained from each departmental manager and cross-checked against individual statements. These cover the approximate percentages of time spent by each person on his main tasks, and are subsequently converted into hours per week. After allowing for additional annual work and overtime, the totals are expressed in hundreds of hours per annum, as shown in the example in Fig. 4.6, the staff utilization table. This table is a means of determining work loads that might be susceptible to diminution and economies in their performance.

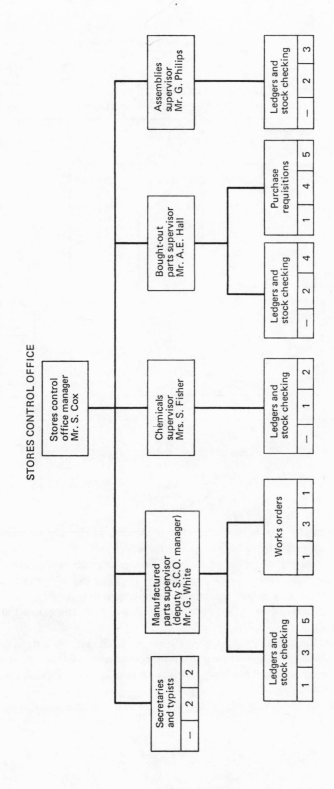

Fig. 4.4 Department organization chart

	NUMBERS OF STAFF			
Grades / Departments	A	B	C	Totals
Stores control office	8	17	22	47
Purchasing	4	9	13	26
Production control	10	25	20	55
Wages	6	10	9	25
Sales invoicing	5	9	18	32
Totals	58	98	114	270

Fig. 4.5 Staff establishment table

1. What arrangements exist for coping with unexpected peaks of work such as are brought about by staff sickness, holidays, and individuals' annual vacations?

2. Are there any avoidable overlaps of work such as the duplication of records or in copying written data from one record to another?

3. What duties are currently performed that are completely outside the scope of a computer—for example, answering telephone inquiries from customers. These tasks, although sometimes occupying a negligible amount of time, can be of vital importance in the functioning of both the existing and the new system.

4. What are the quantities and types of machines used, with their degree of utilization? These include keyboard accounting machines, adding machines, typewriters, punched card machines, copying machines, and any other machines that process data.

5. What skills are possessed by members of staff outside their normal

STORES CONTROL OFFICE	NORMAL HOURS PER WEEK				Totals in 100's of hours per annum inc. additional work
Grades / Routine	A	B	C	Totals	
Filing	—	35	100	135	68
Docket sorting	—	—	105	105	52
Ledger entering	10	160	201	371	190
Stock checking	15	102	122	239	139
Purchase requisitions	26	129	120	275	140
Work orders	20	105	32	157	78
Typing	—	40	65	105	54
Other work	90	63	85	238	146
Idle time	—	10	30	40	17
Totals	320	680	880	1880	968

Fig. 4.6 Staff utilization table

duties? These may have been obtained elsewhere and not officially recorded in any way. They include the ability to operate office machines, a special aptitude with figures, a knowledge of particular systems and techniques, and especially any knowledge connected with computers, data processing, and allied subjects. By discovering these latent skills during the course of the investigation, the systems analyst can give them consideration when the data processing staff members are being recruited (Section 13.5).

4.8 ESTIMATING EXISTING SYSTEM COSTS

An important part of a systems investigation consists of estimating the costs incurred by the existing system, covering all areas of work that might be transferred to the new system. The costs should be subdivided between routines, cost heads, and departments in such a way that comparison with the new system is facilitated later. Not only the present costs of the existing system but also those projected into the future are of interest. Cost changes are likely because of the workload as well as new methods and the increasing cost of skilled clerical labor.

Although an analysis of each department's costs is usually available from the company's costing department, the figures therein may not be entirely suitable for purposes of comparison because of the dissimilarity of cost bases. It is not sufficient to compare blanket departmental costs with blanket computer costs because in reality, whole departments may not be eliminated. Allowance must be provided for continuance of all or some of the duties of the departmental staff within the new system.

A similar argument applies to the usage of office machines. Where the apportionment of machine costs depends upon their utilization in several routines, care must be taken not to assume that this cost will be saved pro rata if one of the routines ceases to employ the machine. The rental or depreciation of machines is not really reduced because their usage diminishes.

The first stage in determining existing system costs is to draw up the staff utilization table as described in the preceding section. It is important that the staff utilization table be suitably split between routines so that these operations can be directly compared with the computer-based routines at a later time. The segregation of staff into grades is somewhat arbitrary and varies from one company to another; grading is a means of reducing the amount of calculation involved and, at the same time, providing reasonably accurate figures. One should guard against having too few persons in the highest grade, since this might again lead to errors if their salaries were apportioned on a pro rata basis.

The departmental costs table, as shown in Fig. 4.7, is constructed from the staff utilization table together with information regarding other costs, such as machines, overhead, and stationery. The wage costs are arrived at by multiplying each grade's total annual hours spent on the routine by the average wage per hour for that grade, and summating the results as shown.

It is not generally worthwhile to analyze material costs except for stationery, the usage of which can be assessed from the results of the study of document movement (Section 4.3). An acceptable apportionment of overhead among routines is achieved by making them proportional to the sum of wages and machine costs for each routine.

STORES CONTROL OFFICE / Routine	COSTS $ PER ANNUM				
	Wages	Machines	Stationery	Overheads	Totals
Filing and sorting	31680	—	840	3168	35688
Ledgers	20688	8088	264	2878	31918
Stock checking	15312	—	24	1531	16867
Purchase reqs	18216	360	154	1858	20588
Works orders	12000	276	101	1226	13603
All other work	6504	156	29	665	7354
Totals	104400	8880	1412	11326	126018

Fig. 4.7 Departmental costs table

4.9 EXERCISES

Problem 1. Verification of Facts

During a systems investigation in a pharmaceutical wholesale house, a systems analyst is told by the chief storekeeper that there are about 13,000 commodities sold from stocks held. How can he verify this figure? It is not, of course, feasible to make a count of the actual commodities in the stores, owing to practical difficulties and the turnover of certain items.

Solution to Problem 1. A figure of 13,000 commodities could be expected to have a tolerance of a thousand or so, either way, so we must not expect exact verification. Possible methods of verification are:

(A) Estimate the number of stores records cards by measuring the space they occupy and calculating from the number per inch, rather than actually counting them all. When doing this, the systems analyst must insure that his measurements are of current stock only and not obsolete commodities.

(B) Examine the last lot of stock-taking records and again make an estimate from the entries per sheet and the number of sheets. It is, of course, possible that such sheets were not used or are no longer available. There should, however, be some records of a similar nature used during the last evaluation of stock by the accounts department.

(C) Examine the purchasing records. These should be held in the buying department for ordering and cost-checking purposes. They may not be too easy to follow, however, owing to their different purpose and their being intermixed with nonstocked commodities.

(D) Read the sales catalog. This should show lists of the commodities sold to customers. Also make sure that the listed items are all currently held and do not include those that are manufactured but not normally held in stock.

Problem 2. Ranges of Items

Domelec, Inc., is a firm wholesaling domestic electrical goods by buying in bulk from the manufacturers and selling in small quantities to retail stores and shops. The firm distributes the goods via a number of warehouses, where it also stores major spare parts.

Suggest what ranges of items are relevant to this situation and what information pertains to each range.

Solution to Problem 2. The main ranges and associated information are tabulated as follows:

Range	Main Information
(a) Goods bought and sold	Commodity code and description, warehouse stocks, purchase price, selling price, sales tax, manufacturer's account number
(b) Spare parts bought and sold	Same as for (a)
(c) Manufacturing suppliers	Manufacturer's account number, name and address, discounts obtained
(d) Retail customers	Customer's account number, name and address, discounts allowed
(e) Staff employed	Employee (check) number, name and address, payroll details

Problem 3. Record Specification Form

The sales department of Domelec holds a master price record (ref. CPR 123) for each of the goods sold. You are to draw up a record specification form (as in Fig. 4.2) to cover the data on the master price record, which is as follows:

The selling price, purchase price, and sales tax are all expressed in dollars and cents; these have maxima of $900, $700, and $300, respectively. Each commodity sold has only one manufacturer, who has been allocated an account number of three digits preceded by a letter from A to F; the highest account number within a letter is 126. The descriptions of the commodities are of varying lengths up to 40 characters, and their codes consist of two letters followed by two digits and sometimes a suffixed letter to indicate a special model. Stocks of a commodity may be held in up to any four of the ten warehouses in quantities from 0 to 250, and this stock figure is amended by the sales clerk when items are sold or received. The warehouses are coded by using the first two letters of their town names.

Solution to Problem 3. This is shown in Fig. 4.8.

Problem 4. Existing Routine Flowcharting

Draw a flowchart to show the operations involved in the "job tickets" routine described below.

Job tickets are collected each morning from the plant office by the cost office, verified, and checked against the control slip with each batch. The plant office is notified of any missing tickets and the remainder are separated into four groups. These are (1) special jobs, (2) normal jobs, (3) maintenance work, and (4) idle time. Special jobs are received by the estimates clerk, who calculates the job cost in conjunction with prequoted amounts. Normal jobs constitute 90 percent of the tickets, and these are entered with the appropriate rates, taken from a rate book, by the cost clerks. Maintenance work is dealt with by one clerk, who enters an agreed rate. Idle time tickets are carefully checked for the plant supervisor's stamp before being passed to a senior cost clerk, who enters a current idle time rate. Unstamped idle time tickets are rejected and sent back to the plant supervisor. All tickets on which rates have been entered are extended by the comptometer section and then passed to the wages department with special job tickets.

Solution to Problem 4. See Fig. 4.9.

		RECORD SPECIFICATION FORM				

NAME OF RECORD Master Price Record				DEPARTMENT Sales		
REFERENCE NO. CPR 123		DATE FILLED IN 9/25/73		FILLED IN BY MDe		

ENTRY REF.	HEADING	PICTURE	MAXIMUM	ENTERED BY	REMARKS
A	Commodity code	AA99	—	O9M dept	
	or	AA99A	—	''	Suffix means special model
B	Commodity descrip	X(40)	X(40)	Sales manager	
C	Purchase price	999.99	$700	Buyer	
D	Selling price	999.99	$900	Sales clerk	
E	Sales tax	999.99	$300	''	
F	Manfr's account	A999	F126	Accounts	
G	Warehouse code	AA	—	Storeman	
H	'' stock	999	250	Sales clerk	
I	'' code	AA	—	Storeman	From 1 to 4
J	'' stock	999	250	Sales clerk	warehouses per
K	'' code	AA	—	Storeman	commodity
L	'' stock	999	250	Sales clerk	
M	'' code	AA	—	Storeman	
N	'' stock	999	250	Sales clerk	

Fig. 4.8 Record specification form of Problem 3

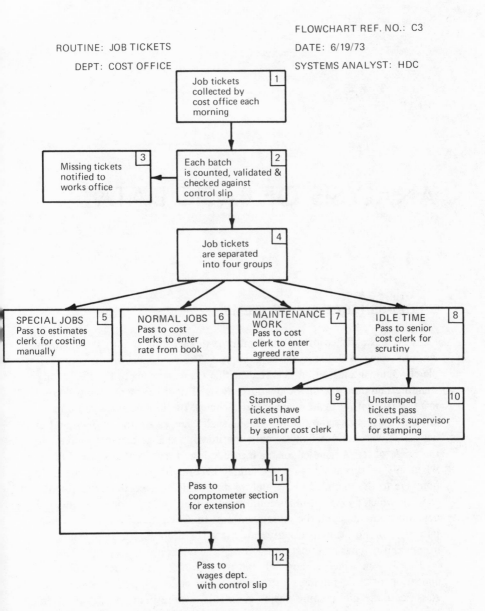

ROUTINE: JOB TICKETS

DEPT: COST OFFICE

FLOWCHART REF. NO.: C3

DATE: 6/19/73

SYSTEMS ANALYST: HDC

1 Job tickets collected by cost office each morning

3 Missing tickets notified to works office

2 Each batch is counted, validated & checked against control slip

4 Job tickets are separated into four groups

5 SPECIAL JOBS Pass to estimates clerk for costing manually

6 NORMAL JOBS Pass to cost clerks to enter rate from book

7 MAINTENANCE WORK Pass to cost clerk to enter agreed rate

8 IDLE TIME Pass to senior cost clerk for scrutiny

9 Stamped tickets have rate entered by senior cost clerk

10 Unstamped tickets pass to works supervisor for stamping

11 Pass to comptometer section for extension

12 Pass to wages dept. with control slip

Fig. 4.9 Flowchart of Problem 4

CHAPTER 5

ANALYSIS OF BASIC DATA

5.1 CLASSIFICATION OF ITEMS

Many well-established office and factory systems function quite satisfactorily without using code numbers of any type. These systems are usually completely manual and can be operated in this manner owing to the staff's long-standing familiarity with the uncoded items. This situation may continue for many years without catastrophe, but a change of circumstances can quickly put an end to this happy state. Among such changes are the employment of new staff who are not familiar with the items, a change or rapid increase in the range of items handled, or the introduction of new methods or machines. When any of these events occurs, it may become necessary to allocate code numbers to the uncoded items. Before doing this, it is worth considering the ways in which a range of items could be classified. The method to be adopted depends primarily upon the subsequent uses to which the classification will be put. It is of no avail to classify raw materials according to their color, for instance, if this has no effect on the use of the materials.

The way in which a range of items is classified will be reflected in the allocation of code numbers to the items. The code numbers encompass the classification, and to some degree can contain more than one classification. However, we must be careful not to overclassify so that the code number becomes too long and complicated.

Changeable Classifications. A range of items may be classified at one point of time into what appears to be unchangeable groups. At a later date, a change of circumstances, perhaps completely beyond the control of the company, may make this classification absolutely valueless. If it is known and accepted that this may happen, the system can be planned so that items will be reclassified whenever necessary, without at the same time disrupting the whole system. The most important point here is to maintain the continuity of identification of items. With this in mind, an identification code should be used for each item in addition to the classifying code.

Methods of Classification

Whatever method of classification is decided upon, it must fulfill all cardinal requirements of a classification system; these are summarized by the following three statements:

1. The method must provide for everything that is to be classified both now and in the future.
2. It must be clear and concise, with readily understood logic—the classification into which an item falls must be immediately apparent.
3. It should not be more specific than is needed for the uses that will be made of it; overcomprehensive methods tend to engender long and complicated code numbers.

Hierarchical Classification. When using this method, items are classified into separate groups according to their most significant characteristic, and are then subclassified within these groups according to another characteristic, and so on. Thus, within a hierarchical list, there is one place only for a particular item.

Suppose, for example, it is necessary to classify a range of raw materials according to three of their main characteristics: substance, price bracket, and form. Using hierarchical classification, this could result in the following arrangement:

Class 1. Valuable Materials

Subclass   1. Nonferrous metals
            1. Sheet
            2. Bar
Subclass   2. Chemicals
            1. In measured containers

Class 2. Moderately Expensive Materials

Subclass   1. Nonferrous metals
            1. Sheet
            2. Bar
Subclass   2. Ferrous metals
            1. Sheet
            2. Bar
Subclass   3. Chemicals
            1. In measured containers
            2. In bulk

Subclass   4. Timber
            1. Sheet (plywood)
            2. Bar (planks)

Class 3. Inexpensive Materials

   Subclass   1. Ferrous metals
              1. Sheet
              2. Bar
   Subclass   2. Chemicals
              1. In measured containers
              2. In bulk
   Subclass   3. Timber
              1. Sheet (plywood)
              2. Bar (planks)

As can be seen from the example, nonexistent items are deliberately excluded; for example, there is no classification for valuable timbers. Using this method, gold bars would be classified as 112, teak planks as 242, and so on. Thus, only 17 different combinations of characteristics are necessary out of the 48 that are theoretically possible in the example.

Faceted Classification. In this method, items are classified according to facets so that each item has a place within every facet. Within each facet there is a further subclassification, which is an amplification of the facet.

Thus, taking the same range of raw materials as above, faceted classification could result in the following codes:

Facet (A) SUBSTANCE   Facet (B) PRICE BRACKET
  1. Nonferrous metals   1. Valuable
  2. Ferrous metals   2. Moderately expensive
  3. Timber   3. Inexpensive
  4. Chemicals

                 Facet (C) FORM
                   1. Sheet
                   2. Bar
                   3. In bulk
                   4. In measured containers

Thus, a material such as gold bars would be 112; sheet steel, 231; bagged cement, 434; and so on. There is always some redundancy with faceted classification because certain combinations, although allowed for, cannot exist, such as valuable ferrous metals, or timber in measured containers.

5.2 CODING OF ITEMS

The use of code numbers as a means of identifying things and persons has become well established within the spheres of business and production. With the growth in variety and complexity of manufactured articles, it has become impossible to identify each item in an unambiguous way except by giving it a unique code number. As a result, many coding systems have been introduced throughout the years, with various degrees of efficiency. The majority of these systems are adequate for data processing, provided they are used correctly and consistently. Coding systems already in use that are unsuitable for a data processing system should be replaced by new coding systems, and in some cases a carefully balanced decision has to be made whether to introduce new coding.

The advantages of a new system from the data processing viewpoint should be weighed against the disadvantage of changeover, with its inevitable problems and the errors that will at first result from it. Changes in even quite small ranges of code numbers can cause much confusion among their users, and it is generally advisable to employ both the old and the new numbers together for some time before phasing out the old numbers.

In computer-based data processing systems, code numbers are the lubricant of the routines, and without them the system would immediately "seize up." If they are badly designed or carelessly used, the system is bound to be inefficient and inaccurate.

Advantages of Code Numbers

What are the particular advantages to be gained from the employment of code numbers within the areas of data processing?

Identification. An item can be identified with absolute precision so that there is no possibility of two different things being confused. These requirement are even more important in a data processing system than in a manual system because there is less chance of the former recognizing an uncoded or erroneously coded item. The computer is not completely helpless in this respect, however, as can be seen in Sections 5.4 and 5.5.

It is very rare for a range of plain language descriptions to be able to provide adequate identification of all items. Even if a precise and unique description were allocated to every item, the tendency of humans to make spelling mistakes would spoil the system's efficiency from the computer aspect.

Storage Space. A code number identifies an item with the employment of far less characters than is needed by a plain language description. Whereas a three-digit code number uniquely identifies a thousand different items, these would usually need names or descriptions of up to eight alphabetic characters. This represents a fivefold increase in computer storage space, and there is also more chance of a misidentification occurring with names and descriptions.

In addition to reducing the computer storage space needed, codes also reduce the space needed on documents, and punched cards and paper tape (and consequently the energy needed to punch them).

Access to Stored Data. As explained in Chapters 7 and 8, code numbers form the keys by which data records on computer files are recognized. They enable the programmer to instruct the computer so that it can give access to individual records in the data base.

Comparison of Items. Similar items are easily recognized by the computer from their code numbers; this facilitates the sorting and summarizing of data so as to provide analyses and totals.

Design of Code Numbers

The classifications of items (Section 5.1) are usually represented by a group of digits placed at the beginning of code numbers, the remainder of the code number consisting of an arbitrary group of consecutive digits. There are, however, many possible arrangements of digits and characters that can be devised. When selecting an arrangement, the systems analyst should always bear in mind the problems of the subsequent users of the codes. The users include the programmers who will later be making a close study of each set of code numbers in order to decide the most suitable methods for accessing records on files.

Before the analyst finalizes a set of code numbers, the prospective users should be given the opportunity of making suggestions regarding the layout. If there is a serious conflict of opinion between the programmers (if already employed at this stage) and other users, the analyst must weight his compromise in favor of the other users. These people will be using the code numbers for an indefinite period in the future, whereas the programmers' problems in this respect are comparatively short-lived.

The most important points to bear in mind when designing code numbers are described below.

Brevity. Code numbers should be as short as possible, particularly if they will be involved in frequent transcription from one document to another. The employment of alphabetic characters furthers this end; three alphabetic characters provide over 17 times the range of three numeric digits (i.e., 17,576 codes as against 1000, with no extra writing effort needed).

Dissimilarity of Content. Where it is desirable to employ alphabetic characters, but the full alphabet is not needed, a range of characters should be chosen based on their dissimilarity to other characters and to the numeric digits. This dissimilarity is, to some extent, a matter of opinion, but a suggested list is

X W L Y F K M P N E A V H Z T R U B D J Q C S G I O

Thus, when a limited range will suffice, it should be chosen from the first part of this list. An eight-character range, for instance, would be X to P, inclusive.

Upper-Case Letters. Computers can normally print only capital letters, so it is best to avoid small letters altogether. In any case, if both types are used in code numbers, human confusion and errors tend to arise. Although certain symbols are printable by computers, these are also best avoided, since they are frequently omitted or misunderstood in clerical procedures.

Similarity of Layout. All the code numbers within a set (i.e., those applying to one range of items) should have similar layouts. This identity of layout encourages both accuracy and completeness in their use. For example, if a set of 700 code numbers is required, it is better to have A100 to A799 than A1 to A700. If it is particularly desirable to start at A1, then nonsignificant zeroes should be introduced (say, A001 to A700) so as to avoid confusion when punching the code into cards.

Gaps within Codes. Gaps between the characters or digits in a code number are better avoided, since they tend to stray and thus cause confusion in punching the code number. When gaps are already in existence, they can be effectively eliminated within the computer by omitting them when punching; in this case they must, however, always be in the same position(s) in each code number so that it can be reverted to the original layout before being printed.

Subcodes. It is often beneficial to build a code number from a group of subcodes. This is very similar to hierarchical classification (Section 5.1) in that it makes the code number partially descriptive and facilitates its use in the preparation of analyses. An example is an employee number consisting of three subcodes:

$$9 \quad\quad 9\ 9 \quad\quad 9\ 9\ 9$$

Factory	Department within factory	Employee within Department

A disadvantage of subcodes is that they can cause a lot of redundancy within a set of code numbers. Taking the preceding example, although there may be only two factories of ten departments each, each department employing 100 persons, a six-digit number is needed to cover these quantities. Thus, although actually there are only 2000 employees in all, the employee number allows for 899,999 (100,000 to 999,999); this represents a redundancy of 99.78 percent in the set of code numbers.

Common Names. When names, descriptions, or addresses contain commonly used names or words, they can be abbreviated by employing code numbers in place of the common names. This technique is especially applicable to large address files such as would hold, among other things, the name of the town in each address. Thus, if an address file refers to, say, 5000 towns among 100,000 addresses, and the average length of a town name is eight characters, the gross saving of storage is 100,000 (8-3) = 500,000 bytes (assuming each town name is replaced by a three-character code). Offset against this is a name look-up table consisting of 5000 entries of 8 + 3 = 11 bytes (i.e., 55,000 bytes). The net saving in storage is therefore 445,000 bytes.

Allocation of Code Numbers

The allocation of code numbers to the items in a range comes after both the classification of the items and the design of the code numbers. Occasionally there is no advantage to be gained from classification, in which case a straightforward consecutive numbering of the items is normally all that is necessary.

Whether classification is employed or not, it is important to insure that:

1. The same code number has not been allocated to two different items.
2. The same item has not been allocated two different code numbers.

The first condition is not difficult to fulfill; this is achieved by making a full list of the allocatable code numbers and then ticking them off when they are allocated. Fulfillment of the second condition may be a far more difficult problem.

The problem of allocating code numbers is alleviated if a range of old code numbers is already in use.

Coding Systems. It is not always necessary to classify every range of items, and indeed in some cases a simple consecutive numbering is all that is required. It is more likely, however, that some more systematic coding of the items is desirable; consequently, one of the methods described below may be adopted. Whatever pertains, it should be remembered that the same code number must not be assigned to two different items, and that the same item must not finish up with two different code numbers. These two cardinal rules, although obvious, are easily contravened by accident when the range being coded is large. The systems analyst can make use of the computer's sorting and comparing capabilities when checking for these types of errors.

Block Codes. A block of consecutive code numbers is allocated to a general class or group of items, allowance being made for expansion of the group. The code numbers have no particular significance as far as individual items are concerned except that they indicate the general class of the items. A typical example of this arrangement would be to give code numbers 1000 to 4600 to manufactured components, 4601 to 5500 to manufactured assemblies, and 5501 to 5900 to manufactured products. A slightly more useful arrangement would be to assign the blocks so that the first digit of the code number indicates the group.

Block coding has the advantages of brevity and simplicity, but it does not provide much meaning or the facility for instant identification of the items concerned.

Interpretive Codes. As the name suggests, these code numbers can be interpreted from the values of their numerals or letters so as to provide a partial specification of the items. This method is useful for a range of items that has a limited number of characteristics because it can incorporate them into the code number. Such a range would be, for instance, for electronic resistors because these have precise measurable characteristics (wattage and resistance) that are conveniently built into code numbers. Code number 3914 would mean that the resistor wattage is 3 watts and its resistance 910,000 ohms, the numeral 4 giving the number of zeroes following the two middle figures.

It is possible to extend this concept to cover wide ranges of items and to specify them in great detail by using the Brisch system, but before embarking upon an ambitious procedure of this nature, it is important to appraise the benefits derivable therefrom.

Mnemonic Codes. When codes have to be remembered by people who do not have reference to manuals or lists, mnemonic coding may prove useful. Although it is unlikely that an entire large range can be remembered in this way, at least partial identification is possible from inspection of the code. A simple variant of this method is to incorporate part of the item's description into its code; for instance, in a set of town codes, BN = BOSTON, NY = NEW YORK, BA = BALTIMORE, and so on.

A different method of employing mnemonic codes is to assign short code names to items without attempting to convey any meaning by the code name itself. This arrangement already applies, with varying degrees of success, to a large number of proprietary articles; for instance, automobiles, drinks, and washing powders. For computer usage it is advantageous if these code names are of fixed length, and for normal purposes they should be as distinct as possible from one another. Mnemonic names may also be considered for limited ranges of items. These are merely groups of letters unique to a given item and selected to facilitate memorizing by making them pronounceable (for example, LODA, NETA). Acronyms, so popular nowadays, such as PERT, COBOL, UNO, and FIFO, are in effect mnemonic names.

5.3 STANDARDIZATION OF INDICATIVE DATA

In this context, "indicative data" refers to all code numbers, descriptions, names, addresses, and similar types of data that are used in data processing systems for other than actual calculations. By standardizing the format and layout of indicative data, the probability of confusion, both within and outside the data processing system, is reduced. As far as is reasonable and practical, standardization should be spread throughout the organization. The implementation of this calls for both discretion and justification on the part of the systems analyst because other departmental staffs naturally resent making what they believe to be trivial and unnecessary changes. In large or widely spread organizations it is advisable to write and distribute a standards booklet; the cooperation of the administration department is valuable in this task.

What things should be standardized? Obviously, the answers to this question depend upon the type of organization or company involved, but the points mentioned below are worth considering.

Week and Day Numbers. Most larger organizations already use week numbers, and this system can be easily extended by numbering the days of the week from 1 to 7. If week numbering is introduced to an organization for the first time, it is usually best to gear the numbers to the organization's financial year.

Twenty-Four-Hour Clock. This obviates any A.M./P.M. confusions on timed events such as computer printouts and time-card "punching."

Dates. It is generally better to employ week and day numbers rather than dates for internal purposes, but where dates (relating to outside transactions) must be used, they should be standardized as numeric digits only: two for the month, two for the day, and two for the last two digits of the year.

In cases where dates are entered on, or read from, foreign documents, care must be taken to insure that the layout is understood. In the United States a date is written month/day/year, but many countries overseas write day/month/year.

Names. These are written and punched in the form of surname followed by initials, with a gap between these and periods between the initials. Separate spaces on documents for surname and initials encourages this format.

Descriptions. If these are logically divisible into parts, the layout of the parts should be standardized when printed on official lists and documents. For example:

BAR,	MILD STEEL	$8 \times 1 \times \frac{1}{4}$ in.
ROD,	BRASS	$10 \times \frac{1}{2}$ in. diameter
TUBE,	COPPER	$7 \times \frac{1}{4}$ in. outside diameter

Trade Customs. Where it is customary to refer to certain things within a trade in a particular way, this should be continued by the new system.

Measurements. Within a given set of items, measurements that are applicable to the items are better specified as all decimal, or alternatively, as all fractional, but not a mixture of both. The choice of units depends mainly upon other factors such as the need for accuracy in the measurements. Where computer calculations are to be carried out using the measurements, decimals are preferable.

5.4 FEASIBILITY CHECKING

Data processing philosophy has become dominated by the feeling that only perfect input data can produce acceptable output results. Although it is obviously desirable to aim for fully correct input data, we live in a real world in which this idealistic target is often unattainable. What measures can be adopted to improve this situation?

The classical methods for checking the preparation of input data are described in Chapter 10, but over and above these methods a series of feasibility checks can be employed to detect errors that may have originated far away from the data processing department. These types of errors include, for example, mistakes made by a customer in his order. Feasibility checks can, of course, be carried out in manual systems; the main difficulties are the time taken and the extra staff needed. The computer is, after all, only doing sophisticated common-sense checks at high speed. Feasibility checks help to identify errors brought about by factors such as faulty clerical work, incorrect punching (although this is also detected by verifying), mistakenly identified batches of data (for example, last week's instead of this week's data), wrong files (also checked by other means), and missing input data.

The degree to which these checks are employed is decided by balancing the level of safeguard needed on the one side against the additional programming necessitated on the other, not forgetting that additional programming occupies both time and computer storage space.

The types of checks that can be employed, either alone or combined, are described in the next few pages.

Limit Checks

Limit checks may be applied to both the input data and the output results from the computer. Each field, or self-contained unit of data, of an input record is checked by the computer program to insure that it lies within certain predefined limits, a similar arrangement applying to output fields. These limits are the maximum and minimum values that the field can normally reach. Examples are: (1) the hours worked in a week by an employee, as stated on his time card, must lie between 20 and 60; (2) the week's earnings of a piece worker, as calculated during the payroll routine, must lie between $100 and $150.

Fragmented Limits.

The field may be predefined as lying within one of a series of separated limits. The field may, for example, be a code number that must lie within one of the ranges 1 to 11, 20 to 28, or 40 to 52. Thus, codes such as 8, 24, and 40 would be accepted, whereas 12, 33, and 53 would not. The series of limits can include zero and infinity; hence, for example, if a field must lie outside the range 100 to 400, it would have two alternative sets of limits imposed on it (0 to 99 or 401 to ∞).

Combination Checks.

A combination check can be applied after two or more fields have been combined in some way or other. This combining is often nothing more than adding or multiplying the fields together; the check is then applied to the sum or product, respectively. *Example:* The limits of purchase quantity are 1 to 1000, and the limits of purchase price are $1 to $90. A purchase order for 500 items at $80 each, although passing the limit checks on the individual fields, could be detected as being invalid by applying a combination check with a maximum of $1000 to the value of the purchase order.

The combination check is equivalent to a limit check on intermediate or final results, and is especially useful for detecting output fields that have become too large for the printing space allocated to them.

Restricted Value Checks

These apply where a field may have one of a short series of individual values and no others. This is really the same as a series of fragmented limits, with each group having its maximum equal to its minimum.

A typical example is to check that the stores number associated with each of the stock transactions is equal to either 1, 2, 5, or 8, because these are the only stores in existence.

Format Checks

Formats are used to check for the presence and relative position of all fields that should be in a record. They are particularly relevant to data read from paper tape, since this medium does not have absolute positions for data as do punched cards (Section 10.2).

Format checking also applies to certain messages in real-time systems; these may not necessarily have their fields in any definite relative order but must nevertheless be checked for completeness. Other messages may contain a mixture of obligatory and optional fields, and so the presence of the former must be confirmed as being among the latter. An example of this occurs in real-time reservation systems, where the obligatory fields are the customer's name, address, and telephone number, and the optional fields are his special requirements (if any).

Layout Checks

The layout of each field is checked for compliance with its proper "picture"; i.e., that each position in the field contains an alphabetic character, numeric digit, or symbol from the range acceptable. *Example:* A part number consists of two alphabetic characters followed by three numeric digits, the acceptable values of which are

1st position	A, B, C, D, G or M
2d position	A to M, or T to W
3d position	1 to 6
4th position	0 to 9
5th position	0 to 9

Thus, part numbers D F 293 and M V 605 are acceptable, but D F 793, E A 538, and A N 123 are not.

A simpler version of this check is merely to confirm that alphabetic positions do not contain numeric digits and vice versa.

Compatibility Checks

The results of two different calculations are checked for compatibility by applying a table of limits or restricted values. This check is best explained by means of an example. Results of practical experience in a factory show that the total machine setup time of a batch of products is roughly related to the batch quantity. Since both of these variables are calculated from data in the men's job tickets, they can be checked for compatibility by using a table such as that following:

BATCH QUANTITY	TOTAL MACHINE SETUP TIME, MIN.
100– 500	20– 30
501–2000	30– 70
2001–7000	60–160

Thus, a total machine setup time of 57 minutes and a batch quantity of 450, both calculated from the same docket, would not be acceptable.

Probability Checks

In the event of failure of any of the checks described in the previous paragraphs, the computer would obey an error program and reject all or some of the related data. There are cases, however, where exact limits are not applicable, but where there is nevertheless a good chance that values outside the limits are erroneous. These probable, but not absolutely certain, errors can be usefully reported by the computer for subsequent human investigation, but should not be rejected at the point of detection.

The example shown in Fig. 5.1 is drawn from the wholesaling of electrical goods, and shows the maximum order quantities to be expected in a single order for a commodity, as related to the commodity group and the class of trade. The probability check would cause the computer to identify excess quantities for human investigation before the goods were actually dispatched.

COMMODITY GROUPS	CLASS OF TRADE			
	1	2	3	4
	Stores	Large shops	Small shops	Clubs and firms
CODES				
10 — 29 heaters, irons	30	10	3	2
30 — 59 radios, record players	10	3	2	1
60 — 89 television sets, washers, cookers	4	1	1	1
90 — 99 lamps, batteries, equipment	1000	100	50	100

Fig. 5.1 Probable maxima of order quantities

Parameters of Feasibility Checks

Owing to changing circumstances, it is quite possible that the limits imposed by a feasibility check at one date will no longer apply at a later date. This situation is handled by treating the limits as parameters; i.e., as things that have to be assigned values prior to each time they are used.

A typical example is the checking of dates on sales tickets during a weekly sales analysis. Since the dates should always relate to the previous week, the parameters are the dates of the first and last days of the previous week.

Flowchart of Feasibility Checks.

A flowchart is shown in Fig. 5.2 and demonstrates how five types of feasibility checks can be applied to an order received from a customer. The details of an order record are:

FIELD	LIMITS	REMARKS
Customer no.	First digit = 1 to 4	First digit is class of trade
Date of order	As per parameters	
Salesman no.	1, 2, 4, 6, 7, or blank	
Commodity code	None	First two digits are commodity group
Order quantity	As per table (Fig. 5.1)	

The feasibility checks applied to the order are:

1. A *limit* check on the date of order, using parameters (boxes 1, 3, and 4).
2. A *restricted value* check on salesman number (boxes 5 and 6).
3. A partial *layout* check on customer number (boxes 6 and 7).
4. A combined *probability* and *compatibility* check applied to order quantity, using the table in Fig. 5.1 (boxes 8 and 9).

5.5 CHECK DIGITS

Check digits are used as a means of detecting errors that sometimes occur when numbers are transcribed from one document or medium to another. It is well known that the human handling of long numbers often leads to confusion, with consequent errors appearing in the numbers. A prime example of this is in the dialing of long telephone numbers, where there is a high level of misdialing.

It has been shown, as a result of various tests,[*] that errors of this nature fall into three main categories:

1. *Transcription errors:* Mistakes in copying a digit of a number (13,795 is copied as 18,795); this category accounts for 86 percent of all errors.

*Beckley, "An Optimum System with Modulus 11," *Computer Bulletin*, Vol. II, No. 3 (December 1967), British Computer Society.

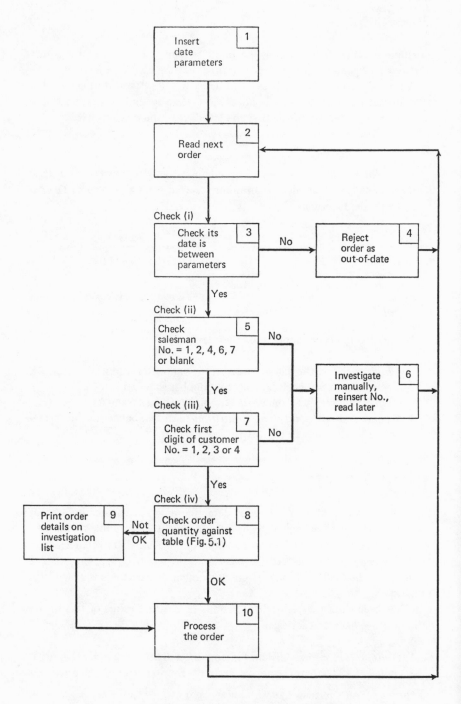

Fig. 5.2 Flowchart of feasibility checks

2. *Transposition errors:* Mistakes caused by swapping adjacent digits of a number (42,691 becomes 46,291); these account for 8 percent of the total errors.
3. *Other errors:* These include the less common errors such as shift, double transposition, omission, and insertion, and account for the remaining 6 percent of the total.

Principle of Check Digits

A method of detecting the errors described above is to use an additional digit (or digits) along with the number itself; this is known as the "check digit(s)." This digit carries no meaningful information other than the assurance of the number's correctness. Its value is related to the rest of the number in such a way that any change in the number is reflected by a change in the check digit.

Prime Number Division. A simple form of check digit can be created by dividing the number by any prime number, and using the remainder as the check digit. For example, the number 2168 divided by the prime number 7 gives a remainder of 5; this could be a check digit and the number always written as 21685. To verify the correctness of a number, it is divided by the prime number (ignoring the check digit), and the remainder from the division compared with the check digit. If these are equal, the number has a very good chance of being correct. If the number had been miswritten as, say, 21645, its division by 7 would result in a remainder of 1, indicating that an error had occurred.

The reason for the choice of a prime number as divisor is that it provides a more uniform spread of remainders than does a nonprime divisor. This happens in practice because ranges of numbers tend to be biased in favor of certain digit values, resulting in a biased (and therefore less efficient) range of check digits if a nonprime divisor were used. This method has certain weaknesses, as is illustrated by the following example: If the number 21685 were miswritten as 28685, a division by 7 would still give a remainder of 5—indicating correctness!

Modulus 11. In order to obtain a higher degree of security, many other methods have been devised. Most of these use a technique involving "weights" and a "modulus." Each digit of the number is weighted (i.e., multiplied by a weight), the results are added together, and their sum divided by the modulus. The remainder from the division is then subtracted from the modulus to obtain the check digit. Because the digits are weighted differently, transcription and transposition errors have a much higher chance of being detected.

A wide variety of weights and moduli have been suggested and supported by statistical arguments. The most prominent system is known as "modulus 11"; this uses a set of consecutive weights and a modulus equal to 11.

The check digit is created as follows:

Step 1. Multiply each digit by its weight; the digit in the least significant
position has a weight of 2; the next position, a weight of 3; and so on.
Step 2. Add together the products of these multiplications.
Step 3. Divide this sum by 11, preserving the remainder.
Step 4. Subtract the remainder of step 3 from 11; the result is the check digit,
but if the remainder is zero, the check digit is also zero.

Example: The code 75264 would be operated upon as below:

Step 1. Code number: 7   5   2   6   4
Weights: 6   5   4   3   2
Products   42 25   8 18   8
Step 2. 42 + 25 + 8 + 18 + 8 = 101.
Step 3. 101÷11 = 9 and 2 remainder.
Step 4. Check digit = 11 - 2 = 9.
Code number is written as 752649.

When a number is to be checked for validity, the check digit is allocated a
weight of 1; steps 1, 2, and 3 are then performed; the remainder should then
be zero.

Example: The code number is now 752649.

Step 1. Code number: 7   5   2   6   4   9
Weights: 6   5   4   3   2   1
Products: 42 25   8 18   8   9
Step 2. 42 + 25 + 8 + 18 + 8 + 9 = 110.
Step 3. 110 ÷ 11 = 10 and 0 remainder.
The code number is therefore valid

An example of the effect of the various categories of errors on modulus 11
numbers is given in Appendix I.

Modulus 10. The method of creating a modulus 10 check digit is as follows:

Step 1. The units position and every alternate digit of the number are multiplied by 2.
Step 2. The digits of the resultant products and the digits of the unmultiplied
positions are all added together.
Step 3. The total is subtracted from the next higher number ending in zero;
the difference is the check digit, unless equal to 10, in which case the
check digit is zero.

Example: The code number 75264 would be operated upon as follows:

Step 1. Code number: 7   5   2   6   4
        Weights: 2   1   2   1   2
        Products: 14   5   4   6   8
Step 2. $1 + 4 + 5 + 4 + 6 + 8 = 28$.
Step 3. Check digit $= 30 - 28 = 2$.
        Code number is written as 752642.

Other Factors. It is possible to use other values for the weights, such as 16, 8, 4, 2, 1, or various arrangements of consecutive numbers; for example, 9, 10, 7, 8, 6, 3, 5, 2, 1. The modulus can also take other values, and these are often the prime numbers just below 100; in the case of 97, for instance, there would be two check digits within the range 00 to 96, giving a higher level of security than a single digit.

If a modulus of 23 were employed, this would give a range of check digits from 00 to 22. These could be conveniently written as A to W, or as A to Z, if G, I, and O were omitted (Section 5.2). A check digit does not have to be situated at the least significant end of the number; it can, in fact, be located in any position within the number. This arrangement is of value when the final real digit of the number has a particular significance in manual procedures and must therefore be easily identified; for example the part numbers of bought-out components end in a 3. A possible difficulty with this arrangement is that the machines creating and checking the check digit may need special modification; this does not apply to computers.

In the case of modulus 11, the check digit is from 0 to 10, and these may sometimes be conveniently represented by the letters A to L (excluding I) or another selection of letters from those mentioned in Section 5.2. If alphabetic characters are undesirable for other reasons and therefore numeric check digits must be used, it is advisable to avoid the value 10 if possible. This value has two digits, whereas the other values have only one, and its use can cause confusion. When creating a new set of code numbers, those with a check digit of 10 should, if possible, be omitted. If check digits are being applied to an existing set of code numbers, unfortunately the 10 must be used. This means that, to avoid confusion, check digits 0 to 9 must be preceded by zeroes and written as 00 to 09; alternatively, the number 10 is written as X.

Desirability of Check Digits

It must be remembered that by associating a check digit with a number, the overall length of the number is increased. This increase will raise the various error rates in addition to making more work in writing the number. The advantages to be gained from the employment of check digits outweigh these disadvantages in the case of long numbers, but for numbers of less than five digits, the advantages are marginal.

Another factor to be taken into consideration is the actual means of generating and verifying the check digits in a business system. There are three main arrangements available:

1. The verification by a computer of check digits written or printed on source documents, and fed into the computer via punching, character, or mark recognition (Section 10.4).
2. The verification by a punching machine of check digits on source documents.
3. Generation of check digits by a punching machine from numbers on source documents, and subsequent verification by computer.

In this context, "punching machine" refers to any paper tape or card punch, or key-operated accounting machine, fitted with a check digit generation or verification unit—these usually employ modulus 11. The stages that are checked by the three arrangements are shown below.

ARRANGEMENT	ENTRY ON SOURCE DOCUMENT	PUNCHING	DATA TRANSMISSION	COMPUTER READER
1	X	X	X	X
2	X	X		
3			X	X

5.6 DETERMINATION OF DATA VOLUMES

During the course of investigating the existing system, a large number of facts are gathered, many of which pertain to data volumes. At a later stage the design considerations applied to the new system will hinge on these volumes. It is therefore wise to give them considerable thought, especially as regards their accuracy and real meaning

Input data volumes are of primary interest, and in particular their degree of fluctuation. The systems analyst's concern is not only for what fluctuation occurs, but also why and when. In some cases the new system will have to deal with input fluctuations as they stand, but in others it is possible to rearrange external events so as to smooth out the input and the consequent work load. Quite trivial circumstances at one point can give rise to marked variations at another. For instance, the inability of the company's postal department to deal with Friday's mail, owing to other special duties, results in this being left over until Monday. This situation, combined with Monday's surge of orders, results in a peak of work for the order department on a Tuesday. With the passage of time, this situation tends to become accepted and regarded as unavoidable, with the inherent danger that it will be perpetuated under the new system.

Another interesting situation occurs when the volume of data has been growing steadily for some time in association with the firm's increasing turnover, or for some other reason. How can the future volumes be predicted? There are, of course, well-known mathematical techniques, such as exponential smoothing,* that forecast future trends from the past figures. Alternatively, a simple graph of previous volumes can be drawn in order to emphasize their trend or cyclic variations. Before acting upon forecasts made by any of these methods, it is prudent to discuss them with the pertinent members of staff. Factors that are neither mathematically nor intuitively predictable may in fact dictate the future volumes. These could be new products, changing competition, different suppliers, new outlets, or other factors. A combination of calculations and opinions, weighted by common sense, gives the best results.

Peak Volumes

All data volumes must be analyzed in order to determine their real effect on the new system; seasonal peaks, such as often occur around Christmas, may decide many of the design points. Before accepting peak volumes at their face value, it is wise to determine their true significance. An example of misleading information in this respect is the statement, "At Christmas there is a 200 percent increase in daily orders received," when further investigation reveals that the number of items per order drops from a normal average of 12 to a Christmas average of only 5. Since it is the volume of ordered items that is of real significance, the increase in work load can be shown to be only 25 percent, as the following figures prove.

	NORMAL	CHRISTMAS	INCREASE
Orders per day	150	450	200%
Average items per order	12	5	
Average items per day	1800	2250	25%

The moral here is always to associate peaks in data volumes with one another and other relevant factors before reaching firm conclusions.

*Fletcher and Clarke, *Management and Mathematics*, Business Books, Ltd., London, 1972.

File Sizes

One of the most important points about existing files is their degree of relevancy to the new system. For example:

1. Is the file data accurate and meaningful?
2. How often is it referred to, and for what reasons?
3. Over what period of time has it been compiled, and how much of its data is now redundant?

The answers to these questions will help the systems analyst to plan the new files so that they have the maximum efficiency combined with the minimum size necessary to achieve this. Manually maintained files tend to grow steadily because they are not limited, except in a very loose way, by the cost of their storage space. As will be seen in the chapters to follow, this is not true of computer files; thus, every effort to eliminate redundant data before the implementation of a data processing system is worthwhile.

5.7 THE NECESSITY FOR SORTING DATA

The amount of sorting really needed is an important factor in the design of a computer-based data processing system, and in some circumstances can be the deciding factor in the choice of the most suitable hardware. The systems analyst must therefore be quite clear regarding the significance and value of any sorting that is carried out in the existing system. It is quite possible that the need for some of this will disappear with the introduction of a computer. On the other hand, certain additional sorting may have to be introduced in order to facilitate achieving the desired results from the data processing system.

There are a good many reasons for the sorting operation in a manual system, and the validity and logic behind these reasons must be established before they are perpetuated in the new system. Some of the main reasons are:

1. To facilitate the manual updating of ledgers by bringing the transaction documents into the same sequence as the ledgers. A typical example of this is the sorting of stores issue and receipt notes into stock number sequence before updating the stock levels of the items on the stock ledger.

2. To accelerate searching a file for data needed frequently, such as prices, stock levels, addresses, and descriptions. Prices, for instance, are needed for billing and ordering; stock levels for purchase control and plant orders; addresses for billing and ordering.
3. To derive sectionalized totals and analyses from sets of documents such as job tickets, bills, and stores notes. This is, of course, the basis of sales analysis by sales area, customer grouping, and so on.
4. To facilitate the manual filing of documents in some particular sequence; for example, purchase orders in order of supplier.
5. To detect exceptions and errors by causing them to appear together on a list. This applies especially to costs and quantities, because sorting into cost or quantity sequence brings together the high or low values at one end or other of the list.
6. To bring together similar or identical items for the purpose of comparison or amalgamation. For example, it might be desirable to compare the sales of a particular commodity group in different years; the relevant sales or order documents are therefore sorted by year within commodity group before being summarized for comparison.

In addition to the reasons given above, there are many other, more minor reasons for the complete or partial sorting of data. If an organization has at some time employed a punched card system, it is possible that the sorting philosophy inherent to such systems will have been continued. This may be true, for instance, when a computer has replaced the punched card system because magnetic tape sorting is effectively a faster replica of punched card sorting.

In general terms, the necessity for sorting in a computer system is inversely related to the size and accessibility of its memory. The ultimate example of this is the human brain; its enormous capacity for facts and rapidity of recall eliminate the need for sorting. In a similar way, a computer equipped with a large amount of direct access storage (see Chapter 7) generally requires less sorting of its input data than does a magnetic tape computer whose access time to its stored data is considerably longer and which therefore must have its input in the same sequence as the stored data that is being accessed.

Whenever a systems analyst comes across a situation where sorting of data is being carried out, he should ascertain the purpose of this with absolute certainty. It is always dangerous to perpetuate sorting without knowing why, since it can be a time-consuming and tedious procedure even when using a computer. On the other hand, if sorting is entirely necessary, understanding its purpose engenders the design of an efficient data processing system.

5.8 EXERCISES

Problem 1. Classification

From the following description of timbers create (A) a hierarchical classification, and (B) a faceted classification, according to their characteristics.

The various types of timber can be segregated into hardwood and soft-
wood, each of which can be further subdivided into durable and nondurable
varieties. Some durable hardwoods are machinable, whereas others are not;
the nondurable hardwoods are dense or lightweight. Only the durable soft-
woods are segregated, and these classifications are construction and non-
construction types.

Solution to Problem 1. The two classifications are selected as follows:

(A) Hierarchical classification
    Class 1. Hardwoods ✓
      Subclass 1. Durable
        Sub-subclass 1. Machinable
        Sub-subclass 2. Nonmachinable
      Subclass 2. Nondurable ⌣
        Sub-subclass 1. Dense
        Sub-subclass 2. Lightweight ✓
    Class 2. Softwoods
      Subclass 1. Durable
        Sub-subclass 1. Construction
        Sub-subclass 2. Nonconstruction
      Subclass 2. Nondurable
(B) Faceted classification
    Facet A: 1 = hardwood, 2 = softwood
         B: 1 = durable, 2 = nondurable
         C: 1 = machinable, 2 = nonmachinable
         D: 1 = dense, 2 = lightweight
         E: 1 = construction, 2 = nonconstruction

Thus, considering three types of timber (i.e., teak, Parana pine, and
obeche), their classifications could be:

TIMBER	HIERARCHICAL	FACETED
Teak	111	11100
Parana pine	211	21001
Obeche	122	12020

The zeroes in the faceted classifications indicate an irrelevant facet.

Problem 2. Interpretive Coding

Design an interpretive code system to cover a standardized range of steel tubes, the lengths of which are 1 through 25 inches at intervals of 1 inch, diameters 0.1 through 3.0 inches at intervals of 0.1 inch, and wall thicknesses 0.02 through 0.08 inch at intervals of 0.01 inch.

Solution to Problem 2. It is preferable to make the wall thickness into the first digit of the code number because it is the only one of the three dimensions that can always be arranged so as to avoid containing zeroes. Therefore, it can be expressed in hundredths of an inch from 2 through 8. Thus, a satisfactory interpretive code number is

DIGIT	DEFINITION
1	Wall thickness, hundredths of an inch
2–3	Length, inches
4–5	Diameter, tenths of an inch

Thus, a tube 1 foot long, 1/2 inch diameter, and wall thickness of 0.03 inch would be coded 31205.

Problem 3. Mnemonic Coding

It is required to form a mnemonic code of three alphabetic characters (letters) for each state of the United States. Each such code must be logically deducible from the name of its state; it must also be unique so as to avoid any confusion, and yet be as mnemonic as possible. (See Appendix I for list of states.)

Solution to Problem 3. The solution is based upon the following considerations.

(A) The first character of the mnemonic code should be the first character of the state's name. This greatly assists in recalling the name, but we must bear in mind that there is a preponderance of certain initial letters such as C, I, M, and N.

(B) The other two characters must make the code unique in spite of the preponderant initials.

(C) Their positions within the name must be deducible by applying simple logical rules, although these rules can vary if necessary between names of different lengths and for double-word names such as NEW JERSEY.

Achieving the given requirements is surprisingly difficult, owing to the commonality of certain letters within the states' names. Quite apart from the initial letters mentioned above, there are many repetitions of the letters A, I, S, and N within the names.

One answer is to adopt the following rules:

TYPE OF NAME	MNEMONIC CODE		
	1st	2d	3d
Five letters or more	1st letter	5th letter	Last letter
Four letters only	1st letter	2d letter	Last letter
Double word	1st letter of 1st word	Last letter of 1st word	1st letter of 2d word
Triple word	1st letter of 1st word	Last letter of 1st word	1st letter of 3d word

Thus:

  California code = C F A
  Michigan code = M I N
  Iowa code = I O A
  New Jersey code = N W J

Problem 4. Feasibility Checking

Suggest what feasibility checks could be applied to the following sets of data if they were to be input to a computer. The type of check(s) and the relevant parameters are required in each case.

(A) The ages of persons at present living in New York.

(B) The heights of American adult males.

(C) Areas in square miles of the states of the United States (see Appendix I).

(D) Overall lengths of American automobiles.

(E) Layout of the Library of Congress catalog card numbers of books published in the United States (see reverse of title page).

(F) Dates of events occurring in 1973 (expressed as mm, dd, yy).

(G) Populations of the states of the United States (see Appendix I).

Solution to Problem 4. The conclusions reached are:

(A) It would not be unreasonable to impose a limit check of 0 through 120 years.

(B) This is more difficult, but a limit check of 4 feet through 7 feet would cover 99 percent or more.

(C) A limit check of 1200 through 600,000, except for District of Columbia.

(D) It is possible that the limits applied could change from time to time, especially at the lower end, but 10 feet through 20 feet would be reasonably satisfactory. The reader can no doubt find more accurate limits.

(E) Layout check: two numeric digits, hyphen, six numeric digits. More research would be needed to decide the precise digit values in each position.

(F) A suitable layout check is:

Digit	Position
1st	0 or 1
2d	0, 1, or 2
3d	0, 1, 2, or 3
4th	0–9
5th	7
6th	3

A more exacting check would be to relate the days to the month (for example, not more than 30 days in September), but this is of marginal benefit.

(G) A limit check of 300,000 through 20,000,000.

Problem 5. Probability Checking

A liquor merchant receives orders from three sources: hotels, clubs, and individuals. His goods are spirits, wines, and beers. Draw up a chart of probability checks that you feel could reasonably be imposed, showing the maximum number of bottles ordered.

Solution to Problem 5. The precise figures to be inserted into the chart are a matter of opinion to a large extent. In reality, they could be based upon this merchant's previous sales figures. However, it is obvious that hotels and clubs order in larger quantities than would individuals, and also that more beer than spirits and wines is sold. A sample solution is:

GOODS	HOTELS	CLUBS	INDIVIDUALS
Spirits	500	1,000	20
Wines	1,500	2,000	50
Beers	5,000	10,000	100

Problem 6. Check Digits

Work out the check digits for the following numbers:

(A) 286057 with modulus 11.
(B) EM 938 with modulus 11 and letters treated as numbers 1 to 26.
(C) 3825794 with modulus 10.
(D) 595724 with modulus 23, weights 2–7, and the check digit expressed as a letter (omitting G, I, and O).

(E) EGEBC (a code drawn from the letters A–G), with modulus 10 expressed as a letter.

Solution to Problem 6. Results are:

(A) 0.
(B) EM938 becomes 513938; check digit is 1.
(C) 5.
(D) Sum of products is 156; when divided by 23, this gives a remainder of 18. This, if subtracted from 23, gives a check letter of E; otherwise we use U (18th letter, excluding G, I, and O).
(E) EGEBC becomes 57523; this gives check letter of C.

CHAPTER 6

DATA PROCESSING FILES

6.1 PURPOSE OF FILES IN DATA PROCESSING

The framework of a data processing system is its files; it is through the proper planning and control of these that the system can function efficiently and comprehensively. The term "file" as used in data processing has a more precise meaning than in manual systems. A data processing file is a collection of data in the form of "records." Each record is discrete and is labeled by a "key" that is held as an identifier within the record. The various arrangements of files, records, and keys (Section 6.2) are legion, and are devised to suit the requirements of a particular system. The set of integrated files used in a data processing system is known as its data base and, as the name suggests, forms the foundation upon which the system functions.

What are the main purposes of data processing files?

1. To hold data in a form that enables it to be processed rapidly by the computer.
2. To make every record accessible to the computer, either individually or en masse; and with a high degree of immediacy in real-time systems.
3. To obtain security and compactness for the data.

These purposes are all achievable, provided suitable hardware is made available for the system and its files are employed in a well-planned manner.

Contents of Files

Virtually any data processed by a computer is held on a file at some time or other; the file may take several forms, including punched cards and paper

93

tape, but here we are mainly concerned with magnetically stored files. The data may be held in a temporary or a permanent state, depending on its type and its purpose. An example of temporary data are the details pertaining to the issue of an item from the factory stores; this data is valuable up to the point at which it amends a stock level or other information, and thereafter becoming increasingly valueless. An example of permanent data is the stock level of an item in the stores; although this level is continually changing, it is held permanently in the file for as long as the item is handled in the factory.

In some applications the distinction between temporary and permanent data is less obvious, but in general terms, temporary data applies to momentary events, and permanent data to reasonably stable facts. Typical examples of temporary data are the details from stores transactions, job tickets, goods delivered, cash received. On permanent files we might well find selling prices, stock levels, manufacturing operations details, and standard costs.

6.2 FILE RECORDS

A data processing file, of whatever type, consists of a relatively large number of records—normally between 100 and 100,000. Each record consists of a number of "fields," one or more of which act as the "key" by which the record is identified. The field(s) that comprise the key may not always be the same within a given set of records from one time to another. The computer is programmed to select the appropriate fields to use as the key during a particular computer run.

An example of a fairly simple record is shown below; there could well be one such record in a file for each item sold over a period of time.

Field A	CUSTOMER NUMBER
Field B	ORDER NUMBER
Field C	WEEK NUMBER
Field D	PRODUCT CODE
Field E	QUANTITY SOLD
Field F	SALES VALUE

The fields A to D are those likely to be used as the key; one computer run may use the customer number, another the product code, and another the week number. These computer runs would, in fact, require only partial identification of the records in order to achieve their results. To identify the records absolutely uniquely, it is necessary in the example given to use all of the first four fields but various combinations of all or some of these four fields may form the key for different runs.

Sorting Keys

As well as being used for identification purposes, a key is also employed as a means of sorting records into a given sequence. These uses are, of course, the same thing because the computer must identify a record before it can sort it. There is, however, a further significance attached to fields when used as sorting keys; this is their relative position when combined into one key. Thus, referring to the preceding record, if we wish to sort a file into customer number within product code, the sorting key would consist of field D followed by field A. If these fields were combined with A before D, the sort would be into product code within customer number.

Fixed and Variable Length Records

The records in a given file are usually of the same type; i.e., they contain similar fields to each other, as in the preceding example. These are known as "fixed length" records because each one occupies exactly the same amount of storage space in the file. There are, however, occasions when it is more convenient to employ records that consist of a variable number of fields. A typical example of a "variable length" record is one that is formed from a customer's complete order; i.e., one containing details of all products ordered on the one occasion. Since all customers do not order the same number of products, the record lengths vary among orders.

Using the same fields as in the previous example, the record now contains:

Field A	CUSTOMER NUMBER	⎫
Field B	ORDER NUMBER	⎬ FIXED LENGTH DATA
Field C	WEEK NUMBER	⎭
Field D1	PRODUCT CODE	⎫
Field E1	QUANTITY SOLD	⎬ FOR FIRST PRODUCT ORDERED
Field F1	SALES VALUE	⎭
Field D2	PRODUCT CODE	⎫
Field E2	QUANTITY SOLD	⎬ FOR SECOND PRODUCT ORDERED
Field F2	SALES VALUE	⎭

Another cause of variable-length records is that they contain variable-length fields. The most common examples of these are in name and address files. The varying lengths of names, streets, and towns combined with the differing number of lines in addresses result in a length ratio of roughly 5:1; i.e., from 25 up to 125 characters per address. It is necessary in this case to insert "end of field" markers between the name, street, town, and county so that the computer can split them up for printing in the conventional manner.

Although variable-length fields may also apply to things such as descriptions of products, it is usually better to treat these as fixed-length fields. This is done by allowing space in every record for the longest description, leaving blank spaces in the field when shorter descriptions apply.

The arrangements given above can result in files of records whose lengths vary considerably and are not necessarily predeterminable. In order that a computer can cope with them, it must be given some means of determining each record's length. This is done by one of two alternative methods.

Record Length Field. This is a field in the record that specifies the length of the record; the unit of measure depends on the type of computer, but can be thought of as the number of characters or words in the record. This field is inserted when the record is created, and amended whenever the record changes in length. This can happen when further fields are inserted or if variable-length fields expand. The "record length" field must be situated in the fixed-length data part of the record so that it can be easily found; often it is the first field in a record.

End of Record Marker. This is a special symbol or group of digits or characters that do not occur elsewhere in the record, and is positioned immediately after the final field in a record. In the example given above, it would be after field Fn if there were n products in the record.

It should be appreciated that variable-length records cannot be sorted into the sequence of fields lying within the variable-length data part of the record. For example, it would not be possible to sort the given records into product code by using D1, D2, etc., as keys because records can in effect be moved only as complete entities when being sorted. In this example the various product codes would conflict with one another with respect to the sequential positions that they should attain. This is not to suggest that variable-length records can never be sorted; sorting, in fact, is quite straightforward, provided there is only one key per record and this is in a fixed position within the record. If necessary, each variable-length record has to be reformed into several fixed-length records before sorting.

Split Records (Spanned Blocking)

If either fixed- or variable-length records turn out to be extremely long, or if there are a few variable-length records that are much longer than the majority, the processing and storage allocation can become inefficient. It is advisable to fix a maximum length beyond which records are split into two or more smaller records. When this is done, the fixed-length data part of the record is repeated identically in each of the smaller records in case it is required for use as a sorting key. It is common practice also to introduce a field containing the "serial number" of the record; this indicates the relationship of the smaller records within the original large record. They are allocated from one upward, unsplit records having a serial number of zero.

Example. A simplified example of split records might apply to our customer orders records.

ORIGINAL RECORD	SPLIT RECORDS
	Serial No. = 1
	Customer No. = 1486
Customer No. = 1486	Order No. = 39521
Order No. = 39521	Product Code = AB43
Product Code = AB43	Quantity = 65
Quantity = 65	
Product Code = BM17	Serial No. = 2
Quantity = 28	Customer No. = 1486
	Order No. = 39521
	Product Code = BM17
	Quantity = 28

6.3 CLASSIFICATION OF FILES

Over a number of years the proliferation of hardware by a wide range of manufacturers has caused an extensive amount of terminology and jargon to be coined. Unfortunately, the various names and terms are not yet completely standardized, with the result that it is easy to become confused. The file terminology used in this edition is, as far as possible, that currently employed by IBM, and where there is a likelihood of confusion, further explanation has been made.

Because of the very wide application of files in data processing, it is worth considering the various ways in which they can be classified. Files are most suitably classified according to their four main facets, and every file falls into one category within each of these faceted classifications at any point in time. As will be realized, a file during processing may change its category to a varying degree within one or more classifications. Thus, the classification of a file is not necessarily static, but can depend upon its operational state during a computer run.

Classification According to Contents

The contents of a file may be considered as falling into one of three categories: movement data, transition data, and master data.

Movement Data (transaction data). These files hold records that have been created from source data and which are a straight copy of the paper tape, punched cards, or other source medium. The records normally contain all data from the source and no other data. A movement file may be sorted into a different sequence from its original, but will still remain a movement file if the contents are unchanged. There is often an unpredictable total of records in the file, and a variable number (including none) having a particular key. For instance, a file of job tickets will vary in size from week to week, and each man may have a different number of tickets.

The contents of movement files are not retained for long, but are replaced periodically by further movement data. It is always possible to reproduce the source data from a movement file by making a straightforward printed copy, after resorting if necessary.

Transition Data. These files generally stem from movement files as a result of changes made to the contents of the latter. The insertion or deletion of fields causes a movement file to become a transition file, the insertions usually being derived from master files. A typical example of this process is the insertion of piecework rates into a job ticket (movement) file from a rates (master) file.

Transition files are not retained permanently and are inclined to be even less permanent than movement files.

Master Data. Master files form a permanent feature in a data processing system, and hold records pertaining to ranges of items of diverse natures. A range is usually semistatic as regards both its size and contents, and there is one record for each item covered by the file. The records may be of fixed or variable length; typical master files with fixed-length records are payroll records, commodity prices, stock levels, personnel records, and with variable-length records are sales ledgers, purchase ledgers, parts operations, product and assembly contents.

A characteristic of master files, which is not applicable to movement and transition files, is that during their life they are frequently "updated," "amended," and "referenced." These procedures are defined as follows:

1. Updating means that regular attention is given to each record in a master file in order to maintain it in an up-to-date condition. This is normally carried out at regular intervals, and it is expected that a substantial proportion of records is updated on each occasion. The process includes, for example, the monthly updating of a sales ledger with invoice and payments data or the weekly updating of the payroll record.

2. Amendment refers to the addition of new records, the deletion of obsolete records, and changes made to existing records that are not expected to occur regularly. The amendment of a master file is not intended to occur at predetermined regular intervals, nor to effect more than a small minority of records, although sometimes practical circumstances override both intentions. An instance of amendment would be the changes made to a price file when new commodities are introduced or when price adjustments are made to existing commodities.

3. Referencing simply means looking up a record so as to obtain data from it without changing it in any way (input mode). An example would be the looking up of the prices of commodities during an invoicing run.

Other terms applied in the updating and amendment processes are *activity or hit rate* and *volatility*. The former is a measure of the proportion of records affected by an updating run or referred to in a referencing run. The latter is a measure of the amount of amendment to which a master file is subjected. It is, however, a somewhat vague term and likely to be confused with activity. However, activity and volatility are somewhat complementary in certain situations such as:

1. *High activity and high volatility:* the work-in-progress file of a machine shop. This occurs because the various jobs being done are constantly updated as the work progresses, and there is also a steady stream of new jobs being added to the file to replace those completed.
2. *High activity and low volatility:* the payroll file of a government department. This assumes, of course, that everyone is paid regularly and that there are comparatively few leavers and newcomers.
3. *Low activity and low volatility:* the customer file of a mail-order firm. Here we have fairly few orders from any one customer but a large number of new customers and of obsolete customers.
4. *Low activity and low volatility:* the inventory file of a museum or art gallery. The activity in this case is zero, since exhibits are not duplicated; the volatility is low because there is little change in the range of exhibits, although this is less prevalent today as a result of easier transportation.

Classification According to Mode of Processing

There are four modes in which files can be processed: input, output, reconstruction, and overlay.

Input Mode (Read Mode). This is the process whereby data is read (referenced or copied) from the file into the core store but is not written back again, the data on the file remaining unchanged. The data is read in differing amounts at one time, depending on the core available to receive it, the amount actually required, and the characteristics of the file hardware.

Output Mode (Write Mode). Data is transferred (or written) from the core store to the file, without having been previously read from the file. This mode is employed during the creation of a file.

Reconstruction Mode. All records in a file are transferred from the file into core store (as in the input mode); after amendment or updating, they are written back into the file, but in other locations. The starting data is preserved in its original locations so that after the run is completed, there are two versions of the file, the opening and the closing file. This process is frequently employed with magnetic tape files, data being read from one reel and written into another. It can also be employed with direct access storage devices as a means of obtaining a high degree of security.

Overlay Mode. Certain records are read from the file into core store and, after amendment or updating, are written back to the original locations in the file. The original data in the file is lost because it is overwritten by the changed records. This process cannot be used with magnetic tape, but direct access storage devices may be employed in this way.

An advantage of this mode over the reconstruction mode is that it is possible to update or amend a small portion of the file without having to process the whole file. Special arrangements have to be made for records that expand during updating and cannot then be accommodated in their original locations. A disadvantage of this mode is its lower level of security, due to the original data being destroyed.

Classification According to Mode of Access and Storage

Mode of access is likely to be confused with mode of storage because the same terminology is employed for both concepts. In some ways, the common terminology is acceptable, since the two modes are closely connected in many cases. Mode of access refers to the way in which the file is referenced, updated, or amended; it is in effect the sequence in which a string of movements is applied to a file. Mode of storage refers to the way in which the file itself is stored, on either magnetic tape or a direct access storage device. In general, the maximum efficiency is obtained when the movements and the file are in the same sequence so that the two sets of records are kept in phase during an updating run, etc. There are, however, many situations that cause the most suitable arrangement to be other than this; a significant factor in this respect is the activity of the file.

The classifications of these two modes are:

> Serial access; serial storage.
> Partitioned access; partitioned storage.
> Sequential access; sequential storage.
> Selective-sequential access; indexed sequential storage.
> Random access; random storage.

Serial Access and Storage. A serial storage file is one in which the records, perhaps having no keys, are stored in adjacent locations so that the file is tightly packed and has no semblance of sequence or other organization.

Typically, a batch of stock movements read from paper tape would form a serial file when written into magnetic tape or disk.

With a serial file it is not possible to deal with isolated records or groups of records; the file must be processed by starting at its beginning and proceeding through to its end. This is to say that it can be accessed only in serial mode; and, moreover, serial access can apply to any of the five modes of storage.

Partitioned Access and Storage. A partitioned file is one consisting of several members, each of which has a unique name or number, and within which the records are stored serially. It is used mainly for the storage of programs and subroutines, or as a core dump (Section 8.3). Partitioned access implies that the computer proceeds directly to the start of a member and then deals with the records in serial access mode. Although magnetic tape can hold a partitioned file (e.g., a library of programs), partitioned access to it is not really possible because all preceding data has to be examined in order to find the wanted member. Partitioned access is quite straightforward on a direct access storage device holding a partitioned file.

Sequential Access and Storage. A sequential file holds its records in a logical sequence of the record keys. The most popular use of magnetic tape is to hold sequential files. Apart from serial storage, this is the only mode of storage with which magnetic tape can be used. Sequential access applies only to sequential or indexed sequential storage; although theoretically possible, it would be very troublesome to deal with a serial or partitioned file in this manner. A random file can be accessed sequentially by a string of sequential movements, but this is in fact tantamount to random access (Section 7.3).

Selective-Sequential Access and Indexed Sequential Storage. In order to save time when looking for a particular record in a sequential file, an index is created that gives the locations of certain records in the file, and the search continues from one of these locations onward. This is known as an indexed sequential file, and can exist only on a direct access storage device. Selective-sequential access is applicable only to indexed sequential storage (Section 7.2).

Random Access and Storage. Care should be taken not to confuse random access and random storage with direct access storage. The latter refers to a hardware device, whereas the first two terms apply to logic concepts.

A random file is characterized by some predictable relationship between the key of a record and the location of that record in the file. The records are otherwise held in a random manner, and so there is no obvious relationship between contiguous records. Random access mode provides the user with the capability of obtaining access to any record directly without having to search through the file. This is particularly relevant to real-time systems, for which the messages are inevitably random.

An interesting point about a random file is that its records cannot be printed in sequence without either sorting them or reading in a sequence of keys. This is because there is no way of ascertaining the key of the next record in the sequence, and it cannot be located without its key being known.

Classification According to Hardware

Files can be stored on a variety of hardware devices; these are chosen to suit the requirements of the particular system, taking other factors as well as file storage into account. The characteristics of the various types of file storage hardware are compared in Section 6.4, but for the moment they can be classified into two main categories: serial access storage and direct access storage devices.

Serial access storage devices are so called because the data on them is accessed and stored only in serial or sequential mode. The only viable form of serial access storage device is magnetic tape, other serial files in the form of punched cards or paper tape being too slow for frequent use.

Direct access storage devices can be used in any of the modes of access and storage mentioned previously. Although there are various degrees of "directness" associated with the different types of direct access storage devices, their main characteristic is the ability to go directly to a stored record without searching through all the preceding records.

Included in this category are devices such as:

> Disk storage drives with removable disk packs.
> Disk storage drives with nonremovable disk packs.
> Magnetic drums.
> Data cell drives.
> Magnetic card files.

Example of Classifying Files. It is required to cost the jobs done in a factory. A contributory factor to each job's cost is the parts used for the job. The parts issued by the stores are therefore entered on stores issue tickets, each of which contains the part number, job number, and quantity issued. These issue tickets are converted into a data processing file (issues file), which is then handled as follows:

1. Sorted into part number sequence.
2. Compared, part number by part number, with another file (price file) that holds all cost prices in part-number sequence; the corresponding part cost price is extracted and multiplied by the quantity issued.
3. A new file is now created for each job ticket (priced ticket file). This contains the part number, job number, and the cost value (price × quantity) of the part for that job.

4. The priced ticket file is used to update another file (job file), which is held randomly.

The various classifications of the above files are:

FILE	CONTENTS	PROCESSING	ACCESS	STORAGE
Issues	Movement	1. Reconstruction*	1. Serial	1. Serial and sequential
		2. Input	2. Sequential	2. Sequential
Price	Master	2. Input	2. Sequential	2. Sequential
Priced ticket	Transition	3. Output	3. Sequential	3. Sequential
		4. Input	4. Sequential	4. Sequential
Job	Master	4. Overlay	4. Sequential (by part no.)	4. Random

*Assuming the file is transferred to another storage area as a result of being sorted. (Numbers refer to items in preceding list.)

6.4 COMPARISON OF FILE STORAGE DEVICES

A considerable number of different file storage devices, covering a wide range of speeds and capacities, is currently available. It is not the intention here to make a fine comparison of the various models, since full technical information about them is readily available from the manufacturers. During the final stages of systems design, the analyst becomes closely involved in making comparisons between competitive equipment, and he will then find it necessary to obtain the latest hardware specifications from the suppliers.

Before this stage, however, the new system has to be designed, at least in broad principle, so as to make it possible to choose from one or more of the categories of file storage devices. These were mentioned in Section 6.3, and can now be further subdivided into three instead of two categories.

1. Serial access devices: magnetic tape drives.
2. Direct access devices with changeable storage units: removable disk storage drives; magnetic card files; data cell drives.
3. Direct access devices with nonchangeable storage units: nonremovable disk storage drives; magnetic drums.

These devices have a multitude of characteristics and to compare them all is beyond the scope of this book. Their main characteristics are, however, worth comparing in order to demonstrate their suitability for a given system.

On-Line Data Volume

This is the volume of data that is accessible to the computer's processor at any one time without human intervention. The devices in categories 1 and 2 above have no limit to the total volume of data that can be stored on the storage medium; i.e., tape reels, removable disks, magnetic cards, or data cells, when off line. There is, however, a limit to the volume that can be on-line at any one time; this limit is imposed by the characteristics of the device and by the number of units of the device fitted to the computer.

Typical on-line data volumes are as follows (note: one byte is one character or two numerals):

Magnetic Tape. From 1 to 100 million bytes, depending on the packing density of the tape and the length of tape per reel. More data can be made available on line by using several tape drives simultaneously, but this is not always a practical proposition.

Removable Disks. Up to 800 million bytes (IBM 3330 with eight disk drives).

Magnetic Card Files. From 340 to 680 million characters (ICL 1958).

Data Cell Drives. Up to 400 million bytes (IBM 2321).

Nonremovable Disks. From 5 to 420 million characters (ICL 2805).

Magnetic Drums. These may have quite small capacities, especially if intended for special-purpose use; the usual range is between 100,000 and 5 million characters.

Access Time

The term "access time" as applied to magnetic tape is not really meaningful because magnetic tape operates in an inherently serial manner. Therefore, any one record is not accessible in a time that is relevant.

Access time is the time taken for the computer to gain access to the data stored on the device and consists for the most part of the time occupied by electromechanical movements of one kind or another.

Removable Disks. These are continuously rotating disks, upon the surfaces of which the data is stored; access time is obtained by read/write heads mounted on movable arms. The access time is made up of head-positioning time and latency; the latter is the delay due to the disk rotation. The average overall access time of removable disks is about 60 milliseconds (IBM 2319).

Magnetic Card Files and Data Cell Drives. These entail the mechanical movement of the cards upon which the data is stored. The drives wrap them around a rotating drum before reading or writing the records. These devices operate best in sequential mode, the random access time being up to about two-thirds of a second.

Nonremovable Disks. The principle of operation is similar to removable disks; their larger size tends to reduce their speeds of operation, giving access times of 3 to 240 milliseconds, depending on size and model.

Magnetic Drums. These are smaller and faster than the devices described above, and there is no movement of the read/write heads. The access time is due entirely to the drum's rotation (i.e., latency), and varies from 0 to 20 milliseconds.

Transfer Rate. This is the rate at which data can be transferred from the storage device into core store, or vice versa. It is of particular importance in relation to the employment of the device for the sorting of records. The approximate transfer rates are:

Magnetic tape: 10,000 to 300,000 bytes per second.
Removable disks: up to 800,000 bytes per second (IBM 3330).
Magnetic card files: 80,000 characters per second (ICL 1958).
Data cell drives: 55,000 bytes per second (IBM 2321).
Nonremovable disks: up to 1.5 million bytes per second (IBM 2305).
Magnetic drums: 50,000 to 1.2 million bytes per second (IBM 2301).

6.5 EXERCISE

Problem 1. Classification of Files

Classify the following files according to (a) content, (b) mode of processing, (c) mode of storage, (d) mode of access, and, in the case of master files, (e) the attention given to them; i.e., updating, amendment, or referencing.

File A. This consists of records with respect to stores issues and receipts, which are being written onto disks in an unsorted manner.
File B. Each record in file A has been read into core store, a cost figure inserted, and the record written back to its previous location so as to form file B.
File C. This is a commodity price file held in commodity code sequence, from which prices are extracted for billing purposes. Each wanted record is found by a search from the beginning of the file, and is then transferred into core store.
File D. A file pertaining to seat reservations on airline flights and which is continually undergoing changes caused by bookings received via online terminals. The records are found by tree searching, and after being brought up to date, are returned to their former locations.
File E. A permanent file of payroll records is held in employee number sequence. Details of new employees are inserted into the file sequentially and the file is written into a new area of storage.

File F. This comprises the results of a public opinion survey and is being read into core a day's group at a time. There is no significant order within each group.

File G. This file holds customers' account details in account number sequence. It is being brought into an up-to-date condition by applying debit and credit transactions. An index is utilized to facilitate finding each account record, and this is subsequently transferred to another area of storage.

File H. This is a permanent file consisting of sales analysis figures held serially in groups. The file is being interrogated by beginning the search for each wanted record from the start of a group and then transferring it into core store after it is located.

Solution to Problem 1. Classifications are tabulated as follows:

File	Contents	Mode of Processing	Mode of Storage	Mode of Access	Attention Given
A	Movement	Output	Serial	Serial	None
B	Transition	Overlay	Serial	Serial	None
C	Master	Input	Sequential	Serial	Referencing
D	Master	Overlay	Random	Random	Updating
E	Master	Reconstruction	Sequential	Sequential	Amendment
F	Movement	Input	Partitioned	Partitioned	None
G	Master	Reconstruction	Indexed sequential	Selective-sequential	Updating
H	Master	Input	Partitioned	Partitioned	Referencing

CHAPTER 7

DIRECT ACCESS FILE TECHNIQUES

The terminology applicable to direct access devices and files is by no means standard. The various manufacturers of this equipment have each introduced their own terminology in their attempts to describe the storage concepts and techniques that can be employed. Unfortunately, this has fostered a wide range of different terms, in some cases using two terms for the same thing and in other cases identifying two things with the same term. This problem is not as bad as it might seem because most computer users are concerned with only one manufacturer's devices, and can concentrate on his terminology. The reader is urged to consult the appropriate manufacturer's manuals in order to discover the precise meaning of the terminology he will be using. For the most part the terminology herein is that of IBM.

7.1 STORAGE CONCEPTS OF DIRECT ACCESS

The storage concepts associated with direct access storage devices are of two kinds: hardware concepts, and software or logic concepts. The former are those connected with the physical electromechanical design of the device. Logic concepts concern the way in which the storage is used to hold data and facilitate access to it. From the systems and programming viewpoints, it is the logic concepts that are the more important, but before describing these in detail a further brief explanation of the hardware concepts is worthwhile.

Hardware Concepts of Disk Devices

The unit on which data is stored consists of a stack of physically connected, rotating disks, on the surfaces of which the data is recorded in the form of magnetized spots. Each recording surface consists of a number of concentric bands, subdivided into blocks as shown in Fig. 7.1. A block is the smallest amount of data that can be transferred to or from the device at one time; in some devices the track must be transferred as a whole, and is therefore equivalent to a block in this respect. It is usual to transfer several blocks at a time into core store, the precise number depending upon the logical layout of storage and the amount of core store available to receive them. Each track on a surface holds identification data in addition to the data for processing.

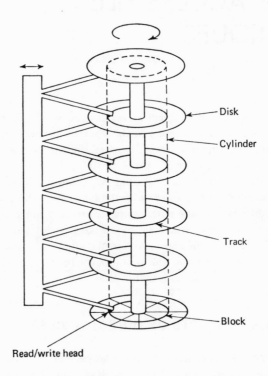

Fig. 7.1 Logic aspects of disk device

Hardware Concepts of Magnetic Card and Data Cell Devices

The storage unit of these devices consists of a number of magazines of magnetic cards, or a number of cells of strips, that can both be loaded by hand into the device. In operation, the required card or strip is selected mechanically and transported to a rotating drum around which it is wrapped. The read/write head(s) is then able to read or write data one track at a time.

Logic Concepts of Direct Access Storage Devices

The systems analyst need not concern himself too deeply with the hardware of direct access storage devices, but should think in terms of three software concepts. These are the cylinder, the bucket (data area), and the record. Records are described in Section 6.2 and are discussed again in this chapter.

Cylinders. From Fig. 7.1 it can be seen that the tracks on the surfaces of the disks comprise a number of imaginary cylinders. The data held in the tracks that form the "surface" of each cylinder can be accessed by the set of read/write heads by positioning them once only. In other words, each track is accessed by one of the heads; the computer then switches electronically to another head and another track is accessed. This process continues through all the tracks until the whole cylinder has been accessed without further head movement.

This concept of cylinders is important because movement of the read/write heads represents a significant portion of the access time, and should therefore be minimized. The cylinder of a data cell drive is one-fifth of a strip (card), since the set of read/write heads can move to one of the five positions.

A magnetic drum cylinder is the complete drum, since normally each track has its own read/write head and therefore the whole of the data on the drum can be assessed without head movement.

Buckets. A bucket (data area) is the unit of data that is transferred with one program instruction. It can consist of one or more blocks, but is normally of constant size within a file. The physical layout of a bucket need not be considered; it is satisfactory to think of it merely as a logical area of storage. During processing, a bucket at a time is transferred into core storage, and it is the amount of core available to accommodate the bucket that largely determines the bucket size.

7.2 SEQUENTIAL PROCESSING

Sequential files are formed by storing records in sequence according to their keys. The records are held within buckets; this means that the records are in sequence within each bucket and that the lowest key in any bucket is greater than the highest key in the preceding bucket. Sequential files are usually processed record after record from the beginning to the end of the file. In this case, there is no difficulty in finding any particular record because the program continues examining each record in turn until the next required record is found. Provided the activity of the file is high and the movements are in the same sequence as the file, this is a satisfactory method. If, however, as is often the case, the file activity is not high, this method is inefficient from a time aspect. A number of other methods are more suitable for low-activity files or for movements not in the same sequence as the file; these are described below.

Self-Addressing (Direct Addressing)

This method is suitable for determining the addresses of fixed-length records in a sequential file where the keys form a complete or almost complete range of consecutive numbers. Under these circumstances, it is possible to find the address of a record by a simple modification to its key.

Suppose, for example, the address is required of the record with key 43746 in a file of 16,000 records, with keys 30001 to 46000. The file is in 2000 buckets, numbered 2001 to 4000, each holding 8 records (Fig. 7.2). The steps to find the address are:

1. Divide the wanted record's key minus the lowest key by the number of records per bucket:

   $(43746 - 30001) \div 8 = 1{,}718$ and 1 remainder

2. Add the first bucket number to the quotient to give the wanted record's bucket; the remainder plus one is the record's position within that bucket:

   $2001 + 1718 = 3719$

Thus the wanted record is the second in bucket number 3719.

This method is not very realistic in that large files with a complete range of consecutive keys are few and far between. It is more likely to be useful for smaller files that might be stored on magnetic drums. Where a file has a limited range of nonvolatile items, it may be worth renumbering them in order to employ this method.

Self-addressing has the following advantages:

1. The key leads directly to the record and so need not be stored within the record. If the key is required for other purposes, it can be easily derived from the address of the record.
2. No index is necessary, thereby saving storage space and searching time.

Self-addressing has the following disadvantages:

1. The records must be of fixed length, although they often are anyway.
2. Nonexistent records whose keys would lie within the range covered by the file must have storage space allocated to them. This is a waste of storage, but may be acceptable if there is plenty of space available.

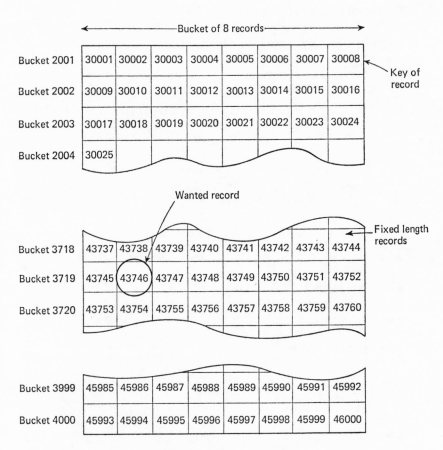

Fig. 7.2 Self-addressed file

Partial Addressing

This method is somewhat similar to self-addressing but differs in that only the approximate bucket number is derived from the key. The wanted record is then found by searching either forward or backward from the beginning of the retrieved bucket. If the wanted record key is greater than the key of the first record in the bucket, the search moves forward, record by record, through the retrieved and succeeding buckets. Otherwise, the search proceeds in the opposite direction.

To be efficient, partial addressing requires a fairly even spread of the keys throughout the range; ideally, records should be of fixed length. The method can be used, however, with variable-length records, but the amount of searching is likely to be greater.

The bucket number of a wanted key is found as follows:

1. Calculate average bucket spread: largest key in file minus smallest key plus 1, divided by number of buckets.
2. Subtract smallest key in range from wanted key.
3. Divide result of step 2 by average bucket spread, ignoring remainder.
4. Add lowest bucket number to result of step 3 to give bucket number of start of search.
5. Search forward or backward from first record in this bucket.

Example. The bucket number of record with key 3157 is required from a file of 500 records with keys in the range 2501 to 3500, stored in buckets numbered 550 to 649, as shown in Fig. 7.3.

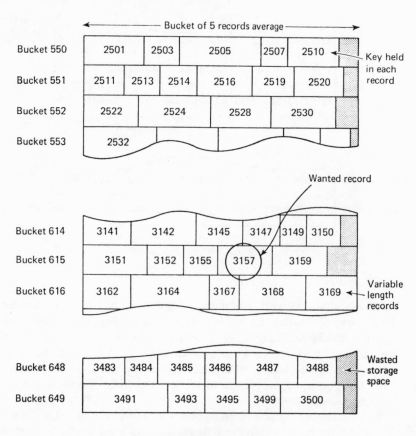

Fig. 7.3 Partial addressing

Proceeding as indicated above,

1. Average bucket spread $= \dfrac{3500 - 2501 + 1}{649 - 550 + 1} = 10$

2. $3157 - 2501 = 656$
3. $656 \div 10 = 65$
4. $550 + 65 = 615 =$ required bucket number for start of search
5. $3157 > 3151$. Therefore, search forward until record is found.

Partial Indexing

This method of locating records is most commonly employed because it is suitable for almost all types of key ranges encountered in practice. It makes use of an index in which are held the highest keys of the records in each bucket. Although, strictly speaking, the records need not be in sequence within each bucket, it is usually more convenient to have them so for practical purposes. Partial indexing corresponds to selective sequential access and indexed sequential storage (Section 6.3). The index described above is known as a bucket index, and it is convenient to hold it in the first bucket of a cylinder, transferring it into core store before use. It may be possible to hold the index permanently in core store if there is sufficient space, such as might well be if only one cylinder is used. If a number of cylinders are in use for one file, a cylinder index is held permanently in core store. This index contains the keys of the highest records within each cylinder, and thereby enables the appropriate cylinder to be found. The bucket index from this cylinder is then used in order to find the wanted record's bucket, and this is searched sequentially to find the wanted record. An example of this procedure is shown diagrammatically in Fig. 7.4, in which it is required to find the record with key 3072.

It will be realized that if the file is being processed against movements that are in the same sequence as the file, the cylinder index is not really needed because each cylinder is used in turn anyway. The same argument also applies to bucket indexes in a file with high activity.

The advantages of partial indexing are as follows:

1. It is straightforward to apply; manufacturer's software is available for creating and searching the indexes.
2. Wasted storage is minimal, up to 100 percent packing density being possible; i.e., available storage utilized (Section 7.4).
3. Variable-length records can be used.

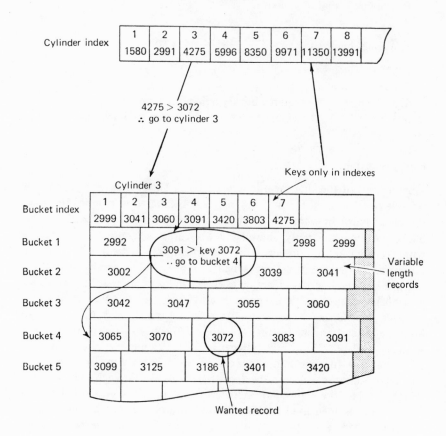

Fig. 7.4 Partial indexing

The disadvantages are:

1. Index tables occupy space in store.
2. Time is taken to search the indexes.

Binary Searching (Logarithmic Searching)

Whereas with partial indexing there is a limited index containing one entry per bucket, binary searching is normally used with a large index containing one entry per record. This entry holds the key of each record so that, having found the wanted record's key in the index, it is possible to go directly to the record. Binary searching can also be used in a modified form with a large partial index so as to reduce the time occupied in searching it.

The principle behind binary searching is to compare the wanted record's key with the key at the midpoint of the index. Depending whether the wanted key is greater or less than the midpoint key, the next comparison is made with the index key situated at the three-quarter or one-quarter point of the index. This principle is continued until identity is found between the index key and the wanted key. This means that each comparison halves the area of the index left for searching; consequently, the number of comparisons has a logarithmic relationship to the size of the index. If the number of entries in the index is E, the maximum number of comparisons is $\log_2 E$, and the average is one less than this. A simple example of binary searching is shown in Fig. 7.5.

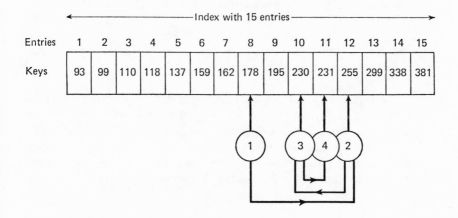

Wanted record's key is 231

1. Compare wanted record's key with midpoint key (entry 8), 231 > 178 ∴ go to three quarter point (entry 12).

2. 231 < 255 ∴ go to midpoint of last two comparisons (entry 10).

3. 231 > 230 ∴ go to midpoint of last two comparisons (entry 11).

4. 231 = 231 ∴ wanted record can be found by using this entry number (11).

Fig. 7.5 Binary searching

It will be noticed in this example that the number of entries in the index is such that it is always possible to halve the remainder at any stage and be left with a whole number; i.e., the number of entries is a power of 2 minus 1. This fortunate state does not usually exist in practice, but it is possible to deal with other numbers of entries by using one of the following methods:

1. Make the number of entries into a suitable amount by inserting dummy entries at the beginning of the index. These must contain dummy keys that are less than the lowest real key; see Fig. 7.6a. It is in fact all right to use these locations for other unrelated data, provided this condition is always fulfilled.
2. Insert dummy entries at the end of the index; these must contain dummy keys whose values are greater than the highest real key; see Fig. 7.6b.
3. A combination of the above two methods, Fig. 7.6c.
4. Modify the binary searching sub-routine so that it ignores "entries" that are outside the limits of the index.

Binary searching is not normally employed with sequential movements unless the file activity is low; it is more suitable for random access to a sequential file. When using a disk device, the index generally resides in core store, but with a magnetic drum the searching is carried out among the records themselves. This is practical with a drum owing to its faster access and the fact that it usually holds a smaller file anyway.

7.3 RANDOM PROCESSING

A random file is one in which the records are stored in such a way that there is no simple relationship between the keys of two adjacent records. As in the case of a sequential file, one or more seek areas are used, and within these are the buckets holding the records. In order to obtain access to a particular record, it is necessary to calculate its address (bucket number) by using its key; this is achieved by employing an "address generation algorithm" (Section 7.4). Random storage is normally applied to files whose activities are comparatively low, and means that the movements need not be sorted. Each movement obtains access directly to the file record with which it is associated by using its key as the input to an address generation algorithm.

Random processing has the following advantages:

1. The movements do not need sorting; since the file is stored randomly, there is no advantage in sorted movements.

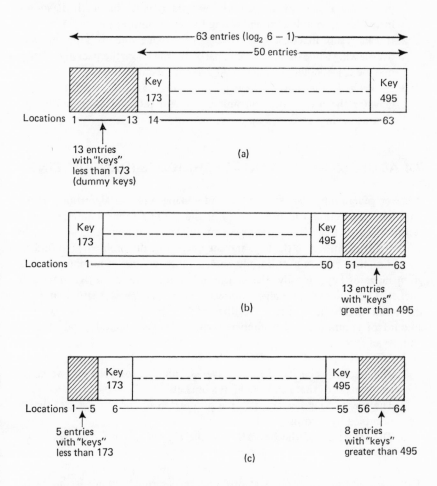

Fig. 7.6 Dummy entries in binary search

2. No index is required, thereby saving storage space and index searching time.
3. The determination of the bucket number by means of an address generation algorithm is generally faster than by retrieving and searching an index. This advantage is of particular importance in real-time systems.

Its disadvantages are:

1. If the file activity is high, this method is less efficient than sequential

processing, except in the case of real-time systems, in which the movements are inevitably random, owing to their manifold sources.

2. It can result in considerable waste of storage due to some buckets overflowing while others are only partially filled; i.e., the packing density may be considerably less than 100 percent (Section 7.6).

3. A change in the range of keys in use, as occurs with a volatile file, aggravates the waste situation unless frequent reorganization of the file (housekeeping) is carried out.

7.4 ADDRESS GENERATION TECHNIQUES (RANDOMIZING)

Address generation, also known as randomizing, uses an algorithm (set of arithmetic operations) in order to form the address (bucket number) of a record from its key. The address generation algorithm involves a few simple arithmetic operations on the key, and the nucleus of the problem is to find an algorithm that creates a uniformly spread range of bucket numbers from a given range of keys. Ideally, the same number of records are assigned to each bucket. Unfortunately, no algorithm can achieve this perfect situation, with the result that some buckets overflow while others are partly empty. The amount of common bucket numbers (synonyms) generated depends upon a number of factors.

1. The bucket capacity; i.e., the average number of fixed- or variable-length records that can be held in a bucket.

2. The file packing density; i.e., the proportion of file storage area that is occupied by records.

3. The structure of the key set, and the algorithm employed to operate upon it.

The main address generation algorithms are described below; it should be borne in mind, however, that these may be combined in order to provide the best arrangement under given conditions.

Prime Number Division

By dividing a key by the number of buckets available, a remainder is formed lying between zero and one less than the number of buckets; this remainder, with a constant added, could be used as the bucket number. However, owing to the bias that tends to be present in a range of keys, the remainders so formed are by no means evenly distributed. This can be seen from Fig. 7.7, in which is a string of keys biased toward ending in a 7. In this example it is

Key	Divisors					
	20	19	18	17	16	15
107	7	12	17	5	11	2
117	17	3	9	15	5	12
127	7	13	1	8	15	7
137	17	4	11	1	9	2
147	7	14	3	11	3	12
157	17	5	13	4	13	7
167	7	15	5	14	7	2
177	17	6	15	7	1	12
187	7	16	7	0	11	7
197	17	7	17	10	5	2
207	7	17	9	3	15	12
217	17	8	1	13	9	7
227	7	18	11	6	3	2
237	17	9	3	16	13	12
247	7	0	13	9	7	7
257	17	10	5	2	1	2
267	7	1	15	12	11	12
277	17	11	7	5	5	7
287	7	2	17	15	15	2
297	17	12	9	8	9	12
307	7	3	1	1	3	7
317	17	13	11	11	13	2
% buckets used	10	100	50	100	50	20

Fig. 7.7 Prime number division of biased keys

desired to assign the records to about 20 buckets, but by dividing the keys by 20, only the remainder 7 and 17 appear out of a possible range of 20 remainders; i.e., only 10 percent utilization of the 20 bucket numbers. This is compensated by using a prime number as the divisor, and it can be seen that 19 and 17 both have a full spread of possible remainders (shown lined). The example does not claim to prove the point, since in practice, keys do not all end with the same digit, but there is often a tendency for this to occur to some extent. If it is known that the bias applies only to the final digit of a key, this can be effectively eliminated by exchanging the positions of the digits before dividing. In practice, however, this is likely to merely bring another biased position into the final position.

In order to determine the extent of bias within keys, it is possible to employ a "digit analysis program" on a computer. This makes a count of each digit value in each position of the key; these counts can then be inspected in order to decide which of the algorithms is the most suitable and which digit positions are best used in the methods described below.

Extraction or Truncation. In this method the most random digits of the key are extracted and formed into a bucket number. If these happen to lie together at one end of the key, the method is called "truncation." *Example:* Key 395628, using first and alternate digit positions, would give bucket number 352; by truncating the last three, the bucket number would be 628.

Folding. The key is split into two or more parts; these are then added together in order to obtain a higher degree of randomness. *Example:* Key 265304, split into two parts, 265 + 304 = 569; split into three parts, 26 + 53 + 4 = 83; using alternate positions, 250 + 634 = 884.

Squaring. The key is squared and a portion of the result used as the bucket number; it is suitable on its own only when the number of buckets available is a power of 10. *Example:* Key 438. 438^2 = 191844. Taking fourth and fifth digits gives a bucket number of 84.

Combinations of Extraction, Folding, and Squaring. The three methods may be combined in any way that increases the randomness of the result. The result still tends to be suited to cases where the number of buckets is a multiple of a power of 10, but this may be overcome by a subsequent division in order to make the final result lie within the required range.

Alphabetic Keys. The address generation algorithms described in the preceding pages have all been concerned with numeric keys, but this does not mean that alphabetic keys cannot be used to form bucket numbers. The general principle is to convert the alphabetic characters in a key into numeric equivalents before injecting them into an algorithm. There is a variety of ways of doing this, and the best method depends not only on the key structure, but also upon the particular computer's internal representation of numerals and letters.

For example, the IBM System/360 eight-bit code representation of alphabetic characters lends itself to using the last four bits only. By so doing, the letters are effectively the same as the digits 0 through 9 because the final four bits of the letters' codes always lie between 0000 and 1001.

7.5 RECORD LOCATING TECHNIQUES

Multikey Accessing

In some applications it is necessary to obtain access to a given record by one of a number of alternative keys. Consider, for example, a stock file in which there is one record per item stocked. Each of these records holds three different unique numbers that might be used as the key for locating it; i.e., part number, bin number, drawing number. The stock file is held on a direct access storage device sequentially by part number because this suits the majority of the processing runs. However, it is necessary in some runs to access the records by using the bin numbers, and in others by using the drawing numbers. This cannot be done simply by attempting to create the required bucket number from either the bin number or the drawing number because the file layout is controlled by the part number only.

The problem can be solved by using address generation algorithms for the bin number and the drawing number that direct the computer to an index location instead of the actual record. Thus, the bin number directs the computer to one location in the index and the drawing number to another (see Fig. 7.8). Both locations contain the bucket number of the record to which the keys pertain. As it is unlikely that the address generation algorithm will provide a unique index location, each location must hold all the bin numbers and drawing numbers that are directed to it, together with the associated bucket numbers. Having found the bucket number, the computer then searches the bucket, examining each record in turn to find the one with the same bin number or drawing number as the key. The drawing numbers and bin numbers could be intermixed in the index, provided there is no possibility of one item's drawing number being the same as another item's bin number.

The example in Fig. 7.9 shows how either bin number 7234 or drawing number JK 93 could be used to locate their record, which is stored in the file in sequence of part number (86639). There are, of course, alternative methods of achieving the same objective, such as employing two indexes—one for the bin numbers, the other for the drawing numbers—held in the appropriate sequence and containing the key and the bucket number within each entry of the index. The appropriate index could be searched by using either a partial indexing method or binary searching.

Tree Searching

This technique is applicable to random files, and causes the records to be stored adjacently when the file is created, thus giving a high packing density. In order to locate a particular record, use is made of a "tree index" that can be held completely or partly in core storage. The index can be regarded as

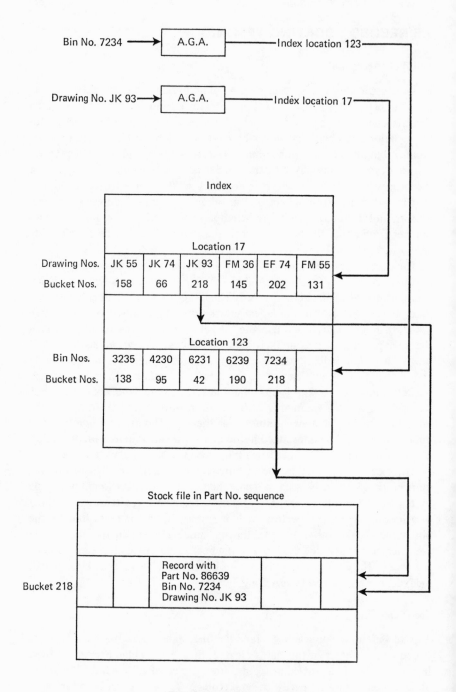

Fig. 7.8 Multikey accessing

122

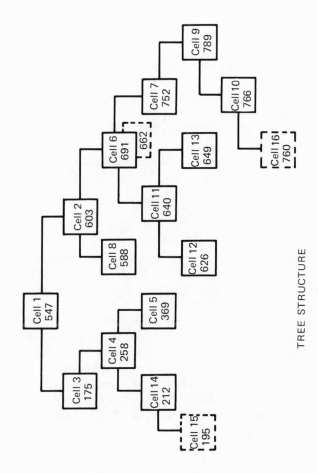

TREE STRUCTURE

Fig. 7.9 Tree searching

Cell No.	Key In Cell	Left Branch Cell No.	Right Branch Cell No.	Deletion Marker
1	547	3	2	
2	603	8	6	
3	175		4	
4	258	14	5	
5	369			
6	691 (662)	11	7	[1]
7	752		9	
8	588			
9	789	10		
10	766	(16)		
11	640	12	13	
12	626			
13	649			
14	212	(15)		(1)
15	(195)			
16	(760)			

TREE INDEX

123

consisting of a number of cells, each of which is directly related to a bucket in the file. The contents of a cell are:

Cell number (can be deduced from its core location).
Key of a record.
Left branch cell number (LBC).
Right branch cell number (RBC).
Deletion marker (DM) (if applicable).

The bucket number is obtained by dividing the cell number by the number of records per bucket (which must be fixed) or alternatively holding the bucket number in the cell.

The principle of tree searching is to compare the wanted record's key with an index key (starting with the first in the index), and then to go to either the left branch cell or the right branch cell, depending on the result of the comparison. This procedure is continued until a comparison gives identity between the wanted record's key and the cell's key. The wanted record can then be found in the bucket derived from the cell number or stored in the cell.

By referring to Fig. 7.9, the procedure for locating a record can be seen. The list of keys shows the order in which they have been put into the file; this order is random according to the key and in sequence of cell number because the cells are taken in turn. As each key arrives, it is put into the first vacant cell, and its cell number is inserted as the left or right branch cell number of the cell from which it branches.

To illustrate the tree-searching technique, let us take as an example a search for key 640 in the index shown in Fig. 7.9:

Step 1. Compare key 640 with the key in cell 1. $640 > 547$. Therefore, move along right branch to cell 2.
Step 2. Compare key 640 with the key in cell 2. $640 > 603$. Therefore, move along right branch to cell 6.
Step 3. Compare key 640 with the key in cell 6. $640 < 691$. Therefore, move along left branch to cell 11.
Step 4. Compare key 640 with the key in cell 11. $640 = 640$. The bucket is found.

Amending a Tree Index. When a record is removed permanently from the file, it is not safe to delete its key from the tree index because this would break the link between the cells in the tree. Instead a "deletion marker" is inserted into the cell so as to indicate that the associated record is deleted; the actual record itself can be either removed immediately or retained on the file temporarily until overwritten by another record.

When a new record is inserted into the file, the cells containing deletion markers are inspected in turn for eligibility to hold the new record's key. This is achieved when (1) all keys branching from the left of the deleted key are less than the new key, and (2) all keys branching from its right are greater than the new key. The new key is inserted into the first eligible cell, the deletion marker is removed from this cell, the record itself is inserted into the appropriate bucket, and the old record is removed. If no cells containing deletion markers are eligible, the key is put into the first vacant cell in the index.

An example of these procedures would be the deletion of records with keys 691 and 789, followed by the insertion of records with keys 662, 195, and 760. Keys 691 and 789 are deleted by searching the index, and when identity of keys is found a deletion marker is inserted into each cell. For key 691 the cells examined are 1, 2, and 6; for key 789, cells 1, 2, 6, 7, and 9. Following this, key 662 is inserted in place of key 691 after also being compared with the keys in cells 1 and 2. It is eligible for cell 6 because this cell contains a deletion marker, and key 662 is greater than all the keys to its left (i.e., 640, 626, and 649) and less than all the keys to its right (i.e., 752, 789, and 766).

For key 760, the cells examined are 1, 2, 6, 7, 9, and 10. Although cell 9 contains a deletion marker, it is ineligible because 760 < 766. Key 760 is therefore inserted into the first vacant cell (i.e., number 16) and a "16" put into the left branch of cell 10 in the index. Key 195 also found the first vacant cell by using the same logic.

Tree searching has the following advantages:

1. High file packing density is possible even with volatile files because deleted records are replaced by new ones.
2. Fairly rapid access to a random file.
3. Records may be fixed or variable length, but if variable, allowance must be made for overflow.

Disadvantages are:

1. A large index is necessary, and arrangements must be made for transferring a section of it at a time into core store.
2. Time is taken in searching the index.

7.6 DISTRIBUTION OF RECORDS

In previous sections of this chapter a number of methods for locating records have been described. These have resulted in the determination of the number of the bucket in which the wanted record should reside. This bucket is known more accurately as the "home bucket" of the record, and is the one to which the record is initially assigned. If, for the reasons that are described below, a record cannot be accommodated in its home bucket, it is known as an "overflow record" and is reassigned to an "overflow bucket."

Sequential File Overflow

When sequential files are created, records are assigned sequentially to buckets and an index is usually created at the same time. No matter whether the records are of fixed or variable length, they are always accommodated at this stage within their home buckets. Subsequently, during file updating and amendment runs, some records expand and others have to be inserted, with the result that the home buckets tend to overflow. The precise causes of bucket overflow in a sequential file are:

1. The insertion of records into buckets that are almost full. This condition can apply to fixed- or variable-length records, except when self-addressing is used with fixed-length records.
2. The expansion of variable-length records. These sometimes have more items inserted during an updating run, with the result that the home bucket can no longer accommodate the expanded record.

Random File Overflow

This occurs for another reason in addition to those mentioned above. It will be remembered that with random files, the home bucket numbers are created by an address generation algorithm. This method always produces synonyms, and any maldistribution of these between home buckets also causes overflow to occur. Thus, with random files, overflow buckets are needed even during the file creation stage.

Overflow Areas

Since bucket overflow is inevitable, some space on a direct access storage device must be allocated for overflow records. With sequential files, the overflow areas are separate from the home areas and may be either (1) within the same cylinders or (2) all concentrated into one cylinder. These are sometimes known as "cylinder overflow areas" and "independent overflow areas," respectively. An advantage of having an independent overflow area is that less space need be reserved for overflow. A disadvantage is that more access time is taken in getting to overflow records, since these are not in the same cylinder as that which is immediately available. A suggested approach is to have cylinder overflow areas that are large enough to contain the average amount of overflow during updating, and to have an independent overflow area for use when a cylinder overflow area is full.

With random files, the overflow area is not separate from the home area because all buckets contain both home and overflow records.

Progressive Overflow

This is the most straightforward but least efficient technique for dealing with overflow records, and is used only with random files. When an overflow occurs, the overflow record is inserted into the next higher bucket that has sufficient space for it. The link between the home bucket and the overflow bucket is simply its nearness. When searching for a record, the home bucket is examined first. If this does not hold it, then successive buckets are examined until the wanted record is found.

Chaining

This method is so called because each home bucket contains a "chaining record"; this is simply a small record at the beginning of each bucket which holds the number of the associated overflow bucket. There are two main types of chaining: progressive chaining and overflow area chaining.

Progressive Chaining. This is very similar to progressive overflow and is used with random files only. The difference is that the chaining record sends the search more directly to the wanted record's bucket. Referring to Fig. 7.10 a search for the record with key A5 with progressive overflow would be via buckets A, B, C, and D; whereas with progressive chaining, the chaining records would direct the search via A, B, and D only.

Overflow Area Chaining. When a record cannot be accommodated in its home bucket, the chaining record of each home bucket directs the search to the first bucket in the overflow area that could hold the wanted record. If this overflow bucket does not in fact hold the record, the search continues from this point onward. This method is illustrated in Fig. 7.11.

With either of the above two chaining methods, expansion of a record beyond the capacity of its bucket (home or overflow) simply entails moving it into the next overflow bucket that can accommodate it. No alteration of the chaining record is necessary, making the procedure straightforward, but chaining can lead to inefficient utilization of storage unless care is taken to reorganize the file regularly. This procedure is known as "housekeeping."

Tagging

This technique is applicable to both sequential and random files, and is more flexible and efficient than the previous methods. As shown in Fig. 7.12, the

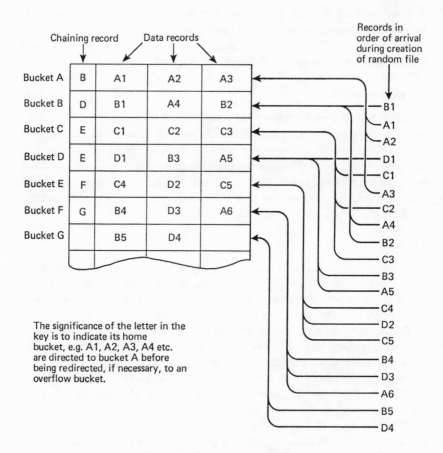

Fig. 7.10 Progressive chaining

principle involved is the replacement of a record that overflows by a "tag" in its home bucket; the tag holds the record's key and the overflow bucket number. Thus, when searching for a record, if a tag is found in its place in the home bucket, the search goes straight to the overflow bucket indicated in the tag.

If an overflow record expands subsequently beyond the capacity of the overflow bucket, it is transferred to the next overflow bucket with available space. The tag in the home bucket is altered to hold the new overflow bucket number; no tag is necessary in the original overflow bucket. The gaps left in buckets by this transfer procedure are filled later by new records or by a

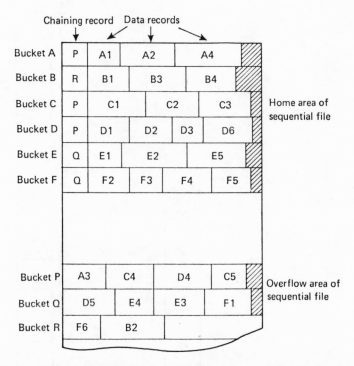

When the file was originally created, records were situated in their home buckets (as indicated by letter in key). During the course of several amendment/updating runs, insertions and expansions of records have resulted in overflow occurring and the reassignment of records as shown. The order in which the records were reassigned was A3, C4, D4, C5, D5, E4, E3, F1, F6, B2.

Fig. 7.11 Overflow area chaining

housekeeping run that not only closes up the file but, in the case of sequential files, completely redistributes the file so as to eliminate tags.

The procedure followed when inserting a record for which the home bucket has insufficient capacity, even for its tag, is to transfer another home record to an overflow bucket and then to insert two tags into the home bucket.

Most of the overflow techniques described and others are supported by manufacturers' software (utility programs and subroutines). The prospective computer user should consult the manufacturers' specifications of these before embarking upon a direct access system.

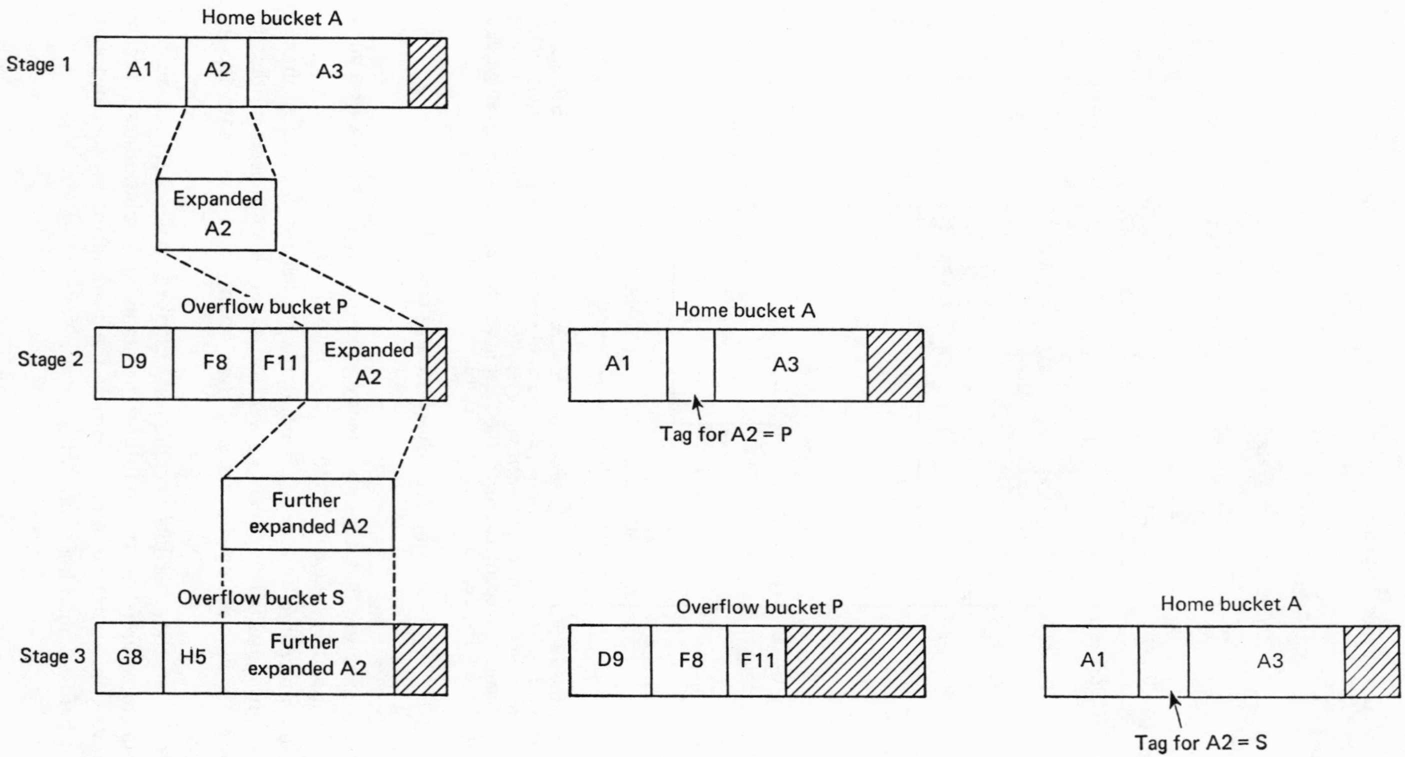

Fig. 7.12 Tagging

130

Assessing the Amount of Overflow

During the updating and amendment of files, new records are inserted, obsolete records deleted, and current records expanded or contracted. In general, with a fairly static file, these changes tend to balance out over a period of time; thus, provided the file is reorganized regularly, there is no need to make a significant allowance for overflow. This is obviously not so with an expanding file, and it is useful to be able to approximate the amount of overflow storage that is needed when the file expands by a known amount. This can be found in Fig. 7.13, which gives a table showing the percentage of

Number of records inserted / Number of buckets in file

	1	2	3	4	5	6	7	8	9	10
0.1	5.0									
0.2	9.5	1.0								
0.3	13.7	1.4								
0.4	17.5	2.0								
0.5	21.4	3.5								
0.6	24.8	4.6								
0.7	28.1	6.0	1.2							
0.8	31.1	7.2	1.5			Less than 1 per cent				
0.9	35.2	8.9	2.0							
1.0	36.8	10.6	2.4							
1.2	41.8	13.6	4.3	1.0						
1.4	46.2	17.0	5.2	1.4						
1.6	50.0	20.4	6.9	2.1						
1.8	53.6	23.6	8.7	2.6						
2.0	56.8	27.0	10.8	3.5						
2.5	63.3	34.8	16.6	6.0	1.7					
3.0	68.3	41.7	22.5	10.9	4.6	1.7				
3.5	72.3	47.6	28.2	15.0	7.1	3.0	1.2			
4.0	75.5	52.7	33.7	19.5	10.2	4.8	2.1			
4.5	78.0	57.2	38.8	24.2	13.8	7.1	3.4	1.4		
5.0	80.1	61.0	43.4	28.7	17.6	9.9	5.1	2.9	1.5	
5.5	81.9	64.2	47.6	32.9	21.4	12.9	7.3	4.1	2.2	
6.0	83.4	67.0	51.4	36.9	25.4	16.2	9.7	5.6	3.1	1.9
6.5	84.7	69.4	54.7	40.7	29.1	19.2	12.1	7.1	4.1	2.2
7.0	85.7	71.5	57.7	44.6	32.8	22.8	14.9	9.2	5.4	2.9
7.5	86.7	73.4	60.2	47.6	35.4	25.9	17.5	11.5	6.9	4.5
8.0	87.5	75.0	62.7	50.7	38.5	29.4	20.9	14.3	9.6	6.6
9.0	88.9	77.8	66.8	55.9	45.4	35.5	26.7	19.2	13.2	8.6
10.0	90.0	80.0	70.0	60.1	50.4	41.1	32.4	24.6	17.9	12.3

Average spare bucket space (records)

Fig. 7.13. Percentage of inserted records assigned to overflow buckets

inserted records that are assigned to overflow buckets. In order to use the table, it is necessary to find the average spare bucket space, ASBS (in records): If

R = total number of records in file
B = number of buckets in file
r = average record size in characters
b = bucket size in characters

then the average spare bucket space is given by

$$\text{ASBS} = [b/r] - (R/B)$$

where the square brackets indicate that this quotient is rounded down.

Example: Suppose we insert another 24,000 records into a file of 100,000 records of average length of 200 characters, stored in 20,000 buckets of 1,500 characters. Then

$$\text{ASBS} = \left(\frac{1500}{200}\right) - \frac{100,000}{20,000} = 7 - 5 = 2 \text{ records}$$

When 24,000 records are inserted,

$$\frac{\text{Number of records inserted}}{\text{Number of buckets in file}} = \frac{24,000}{20,000} = 1.2$$

Using the table in Fig. 7.13, we find that 13.6 percent (i.e., 3264 records) can be expected to overflow. This is an interesting discovery because, at first sight, one might assume that as there is plenty of spare storage space, no overflow would occur.

Total storage space = 20,000 buckets X 1,500 characters = 30,000,000 characters
Overall file size initially = 100,000 records X 200 characters = 20,000,000 characters
Spare storage space = 10,000,000 characters
Insertions = 24,000 records X 200 characters = 4,800,000 characters

Thus, in spite of the fact that only 48 percent of the spare storage space is being taken up, overflow occurs.

Random File Distribution

It will be recalled from Section 7.4 that an address generation algorithm is an attempt to operate on a set of keys in order to obtain an even spread of bucket numbers. The degree of success achieved depends upon both the algorithm and the structure of the set of keys. With a random file, the algorithm is used not only to retrieve records during processing runs, but also to assign them to buckets when the file is being created. This means that there is the possibility of overflow occurring during file creation because, although the algorithm might be perfect for an even spread of keys, in reality this is not often the case. It can be shown from probability theory that the expected number of buckets b_r that will have r records assigned to them is given by the formula

$$b_r = \frac{B \exp[-R/B]}{r!} \left(\frac{R}{B}\right)^r$$

where B is the total number of buckets in the file and R is the total number of records in the file.

This is made clearer by using an example. Suppose there are 20,000 records to be stored randomly in 4000 buckets; then

$$\frac{R}{B} = \frac{20,000}{4000} = 5$$

and

$$B \exp[-R/B] = 4000 \times 2.72^{-5} = 26.96$$

Thus, for instance, the number of buckets with two records assigned is

$$26 \cdot 96 \times 5^2/2! = 337$$

In a similar way, the number of buckets with other numbers of records assigned can be calculated and put in the form of a table, as shown in Fig. 7.14. These figures have been rounded off to whole numbers, which accounts for the small discrepancies in the totals.

Suppose the buckets are large enough to accommodate up to seven average records; this corresponds to an approximate packing density of

$$20,000/4000 \times 7 = 71 \text{ percent}$$

We can now calculate the number of records overflowing. Looking at the cumulative totals at the seven records assigned level, we see that the first 3468 buckets have 15,249 records assigned to them. This means that the remaining 532 buckets are all full because an attempt is made to assign eight

Number of buckets	Number of records assigned	Cumulative total buckets	Cumulative total records
27	0	27	0
135	1	162	135
337	2	499	809
562	3	1,061	2,495
702	4	1,763	5,303
702	5	2,465	8,813
585	6	3,050	12,323
418	7	3,468	15,249
261	8	3,729	17,337
145	9	3,874	18,642
73	10	3,947	19,372
33	11	3,980	19,735
14	12	3,994	19,903
5	13	3,999	19,968
2	14	4,001	19,996
1	15	4,002	20,011

Fig. 7.14 Records assigned to buckets

or more records to each one; thus they hold seven records each (i.e., 532 X 7 = 3724 records between them). The total number of records stored in their home buckets is 15,249 + 3724 = 18,793, leaving 1027 to be stored in overflow buckets (i.e., 5.1 percent).

This figure can be confirmed by referring to Fig. 7.15, which shows the percentage overflows for various packing densities and bucket sizes.

7.7 SECURITY OF DIRECT ACCESS FILES

It is an unfortunate fact of life that no equipment, computer program, or computer staff are infallible. This means that every system must incorporate measures to insure that no data is permanently lost. This safeguard is most important of all for master files because these may have been updated and amended many hundreds of times since they were originally created, and it is therefore out of the question to recreate them from the original source data.

The ways in which master file data can be lost from direct access files are mainly as follows:

1. Incorrect movements data is used to update the master file. This should obviously be detected by the system before damage is done.
2. A programming or operational error results in the accidental overwriting

Bucket size in records	Packing density $= \dfrac{\text{Number of records}}{\text{Storage capacity in records}}$					
	50%	60%	70%	80%	90%	100%
1	21.3	24.8	28.1	31.2	34.1	36.8
2	10.4	13.7	17.0	20.4	23.8	27.1
3	6.0	8.8	12.0	15.4	18.9	22.4
4	3.8	6.2	9.1	12.3	15.9	19.5
5	2.5	4.5	7.1	10.3	13.8	17.6
6	1.7	3.4	5.8	8.8	12.2	16.1
7	1.2	2.6	4.7	7.6	11.0	14.9
8	0.8	2.0	4.0	6.7	10.1	14.0
9	0.6	1.6	3.4	5.9	9.3	13.2
10	0.4	1.3	2.9	5.3	8.6	12.5
12	0.2	0.9	2.2	4.4	7.5	11.4
14	0.1	0.6	1.7	3.6	6.7	10.6
16	0.1	0.4	1.3	3.1	6.0	9.9
18	0.1	0.3	1.0	2.7	5.5	9.4
20		0.2	0.8	2.3	5.0	8.9
25		0.1	0.5	1.7	4.1	8.0
30			0.3	1.2	3.5	7.3
40			0.1	0.7	2.6	6.3
50			0.1	0.5	2.0	5.6
70				0.2	1.4	4.8
100				0.1	0.8	4.0

Fig. 7.15 Percentages of initial records overflowing in a random file

of master data; very probably, the overwriting data has no connection whatsoever with the master data.

3. Damage to removable disks during storage or transit; this might be physical or electrical (magnetic) damage.
4. Breakdown or temporary failure of equipment; the latter could be caused by mains interference, power failure, or unsuppressed radiation from other equipment.

With sophisticated systems, the recovery procedure after the detection of an error is automatic and independent of the particular program(s) in operation at the time. This must be so with real-time and multiprogramming systems, because the operators cannot be expected to trace manually the faults in such complex systems. With more mundane systems, however, file security has to be arranged by the user. A few methods are described below.

File Copying

This involves copying (dumping) updated direct access files onto magnetic tape or another area of direct access storage. The precise method is dictated by the configuration of the computer. In some cases, especially with fixed disk storage devices, magnetic tape is included mainly for security purposes. When doing a long updating run, the runs are split into sections, and a dump is made after the completion of each section. This enables a restart to be made at the beginning of the section in which trouble occurred, rather than restarting at the beginning of the entire file.

After-State Copying

During an updating or amendment run, this method calls for the buckets that have had changes made to a record therein to be written into another file. This occurs every time a bucket's records are changed, so that at the end of the run or after several runs, there are several "after-state" versions of each bucket. These are then sorted and merged to produce the most recent version of each bucket, and these are merged into the original file as it is being copied to produce the updated file.

A master file is thus overlayed far less frequently than usual, with a corresponding reduction in accidents.

Record Deletion

Upon occasions some of the records in a file become obsolete and serve no purpose by remaining in the file. Once the user is certain that no further information is extractable from them, they can be deleted from the file. In some applications, this is an automatic process; for example, where an address file of transient customers is in use. If the date is inserted into the record by the computer every time it looks up the address, it becomes a straightforward procedure for a housekeeping run to delete records that have not been referred to for, say, over one year.

As a general principle, a printed copy should be made of every record deleted from a file. This serves two purposes: first, the copy can be retained for possible reference; second, it is used to check that all the intended records have been deleted, and no others. This check is especially necessary when the deletions are based on human decisions. Each record will be found and deleted by the use of a key, such as the customer's account number in an address file. It is vital that no mistake be made in the key used. By printing the name and address a further human check can be made to make sure that proper deletion has been effected.

An extension of this type of check is for the computer not to delete an obsolete record immediately but to insert a deletion marker into it when printing the hard copy. During the next housekeeping run, all marked records are automatically deleted. If, in the meantime, an error is found to have occurred, the deletion marker is removed before the housekeeping run. This method avoids having to reinsert records that have been deleted erroneously.

7.8 EXERCISES

Problem 1. Self-Addressing

Describe how to organize a self-addressed sequential file of 12,000 fixed-length records whose keys lie within the range 25000 through 37999. Each record in the file contains 160 bytes and the buckets contain 1000 bytes; the first bucket in the file is number 501.

What would be the addresses of the records with keys 27932, 36057, 31598?

Solution to Problem 1. Since the buckets have 1000 bytes and the records have 160 bytes, it is possible to contain six records in each bucket with 40 bytes of wasted storage. The range of keys covers 13,000 records, although there are actually only 12,000 in existence. Thus, the file could be stored in buckets 501 and onward, actually occupying

$$13,000 \div 6 = 2166\frac{2}{3} \text{ buckets}$$

or up to and including bucket number 2667.

$$\text{Packing density of file} = \frac{12,000 \times 160}{2167 \times 1000} = 88 \text{ percent}$$

$$\text{Address of record with key } 27932 = \frac{27,932 - 25,000}{6} = 488 \text{ with 4 remaining}$$

$$501 + 488 = 989 \quad \text{and} \quad 4 + 1 = 5$$

The record is therefore the fifth in bucket number 989.

Similarly, 36057 is the sixth in bucket 2343, and 31598 is the fifth in bucket 1600.

Problem 2. Partial Indexing

Create a selective sequential file consisting of 20,000 variable-length records of from 50 to 150 bytes, averaging 90 bytes. Buckets of 2000 bytes, starting at number 1600, are available for storing the data and index. Describe how the file and index could be arranged if the records' keys occupy 3 bytes each; partial indexing is to be employed, and the complete file can be accommodated in one cylinder.

Solution to Problem 2. Because the records are of variable length, we cannot be absolutely precise with regard to the file's layout, but in general terms it could be as follows:

$$\text{Average number of records per bucket} = 2000 \div 90 = 22$$

Thus,

$$\text{Number of buckets needed} = 20,000 \div 22 = 909$$

It follows that there are 909 keys in the index, and therefore this will occupy

$$909 \times 3 = 2727 \text{ bytes}$$

or, effectively, two buckets. The index will therefore occupy the whole of bucket number 1600 and approximately the first third of bucket 1601. The records will occupy buckets 1602 through 2510, approximately, and

$$\text{Packing density of file} = \frac{20,000 \times 90}{909 \times 2000} = 99 \text{ percent}$$

Problem 3. Extraction or Truncation

Shown below is a representative sample from a set of keys applicable to records in a randomly stored file. Determine the best digits to extract as bucket numbers 00 through 99 so as to obtain the most uniform distribution of records between buckets.

28605	19537	82861	38955	26244
53052	20470	31980	45463	10382
19567	26964	14666	90936	42062
69017	17466	74728	33071	32208
41767	34803	22059	26575	48958
15374	59918	17520	52549	15549

Solution to Problem 3. A digit analysis of the keys given above provides totals under

	1st digit	2d digit	3d digit	4th digit	5th digit
0	0	3	5	3	3
1	8	2	0	2	2
2	6	5	2	2	3
3	5	2	2	2	2
4	4	3	3	3	3
5	3	3	6	4	3
6	1	3	2	8	3
7	1	2	2	4	4
8	1	3	2	2	4
9	1	4	6	0	3

From this analysis it is readily apparent that the second and fifth digits are the most random and will therefore give the most even distribution. It should be borne in mind that the keys in the problem are only a small sample and insufficient to truly illustrate this uniform distribution if used to form bucket numbers.

Problem 4. Tree Searching

(A) Create a tree index to contain the keys below in the order shown.

520, 697, 185, 404, 484, 616, 591, 435, 900,
862, 455, 103, 576, 699, 750, 247, 388, 501.

(B) Amend the index to show its contents after
(a) 404, 616, 900, and 576 are removed, and
(b) 555, 592, 833, 850, and 370 are inserted.

Solution to Problem 4(a) and (b). Indexes and amendments are listed below. Bracketed figures are those inserted as a result of the amendments.

CELL NO.	KEY IN CELL	LBC	RBC	DM
1	520	3	2	
2	697	6	9	
3	185	12	4	
4	404	16	5	[1]
5	484	8	18	
6	616[592]	7		
7	591	13		

(Continued)

(Continued)

8	435		11
9	900	10	[1]
10	862	14	
11	455		
12	103		
13	576[555]		
14	699		15
15	750		[19]
16	247		17
17	388	[21]	
18	501		
19	[833]		[20]
20	[850]		
21	[370]		

Problem 5. Overflow Area Chaining

The records with keys as shown below are to be assigned to home buckets 1000 through 1009 by extracting the third digit of the key and adding 1000 to form the bucket number. All buckets are 1000 bytes, and those records failing to be accommodated in their respective home buckets are to be assigned instead to buckets 2000 and onward (the overflow area).

Show what the situation will be with regard to assignments after all the records have been dealt with. The chaining record occupies 1 byte.

INPUT ORDER	KEY	BYTES	INPUT ORDER	KEY	BYTES
1	5611	200	16	4706	250
2	5596	150	17	5813	200
3	5275	250	18	3760	250
4	5618	200	19	4295	200
5	5347	400	20	5030	400
6	4792	300	21	4060	250
7	5820	250	22	4562	100
8	5155	350	23	3823	100
9	4860	300	24	5768	50
10	4598	400	25	5519	150
11	5037	200	26	5298	200
12	5146	300	27	5360	400
13	5217	250	28	3745	350
14	4852	200	29	3890	200
15	5193	100	30	4634	300

Solution to Problem 5. Reassignment of records is tabulated as follows:

BUCKET NO.	CHAINING RECORD	KEYS OF RECORDS IN BUCKET					UNUSED BYTES
1000		4706					750
1001	2000	5611	5618	5217	5813		49
1002		5820	3823				650
1003		5037	5030	4634			100
1004	2001	5347	5146				299
1005		5155	4852				450
1006	2000	4860	3760	4060	4562	5768	49
1007		5275					750
1008							1000
1009	2000	5596	4792	4598	5193		49
2000		4295	5519	5298	5360		50
2001		3745	3890				450

Problem 6. Random File Distribution

Using the table in Fig. 7.15, determine the absolute minimum number of overflow buckets that should be initially allocated for each of the following random files:

(a) 10,000 records of 600 bytes each, assigned to 2500 home buckets of 3000 bytes each.

(b) 20,000 variable-length records of from 50 to 100 bytes assigned to 1000 home buckets of 2000 bytes. The average record length is 90 bytes.

(c) 50,000 records of 30 bytes each, assigned to 1500 home buckets of 1000 bytes each.

Solution to Problem 6. Allocation of overflow buckets is as follows:

(a) Bucket size in records = 3000 ÷ 600 = 5.

$$\text{Packing density} = \frac{10,000}{2500 \times 5} = 80 \text{ percent in table}$$

Actually, this is 72 percent when taking overflow space into account. From the table,

$$\text{Initial overflow} = 10.3 \text{ percent}$$

or 1030 records. Then

$$\text{Overflow buckets} = \frac{1030}{5} = 206$$

(b) Bucket size = 2000 ÷ 90 = 22 records.

Packing density = $\dfrac{20,000}{1000 \times 22}$ = 90 percent in table (actually 86 percent)

From table,

Initial overflow = 4.6 percent (interpolating)

Or 920 records. Thus,

$$\text{Overflow buckets} = \frac{920}{22} = 42$$

(c) Bucket size = 1000 ÷ 30 = 33 records.

Packing density = $\dfrac{50,000}{1500 \times 33}$ = 100 percent in table (actually 93 percent)

From table,

Initial overflow = 7.0 percent

Or 3500 records. Then

$$\text{Overflow buckets} = \frac{3500}{33} = 106$$

CHAPTER **8**

MAGNETIC TAPE FILES

Magnetic tape is a widely used storage medium for batch processing applications, in which computer routines take a batch of input data (usually in the form of paper tape or punched cards), process it, and produce a discrete amount of output information. Magnetic tape is in many ways eminently suitable for the conventional data processing of accounting, costing, and stock control procedures. This is especially true where these applications were carried out previously on punched card machines. During the period 1960–1965, most computers installed were equipped with magnetic tape units, and their users saw them as extensions to punched card systems, with the magnetic tape replacing the files of punched cards.

It is worth recalling the advantages of this medium and remembering that they also apply, to a different degree, to direct access storage devices.

1. A reel of magnetic tape can hold several million characters, and is therefore equivalent to several hundred thousand punched cards.
2. Data is transferred to or from the tape at speeds of 10,000 to 300,000 bytes per second. This facilitates the sorting of data records at high speed, and the matching and merging of records from different files.
3. Data can be stored indefinitely or erased and new data recorded on the same reel. Since reels of magnetic tape are comparatively cheap as compared with magnetic disks, it is feasible to hold a large amount of data off line in the tape library.
4. The length and content of data records are extremely flexible, more so than with direct access storage. This facilitates the updating, amendment, and housekeeping of files from the programming aspect.

5. Source data can be keyed directly onto magnetic tape. Thus the cost of punching media is avoided, although the initial capital outlay is greater.

The main disadvantage of magnetic tape, particularly for the future, is its inherent characteristic of serial storage. This obviously limits its ability to provide immediate access to stored data, with consequent problems in real-time and conversational mode applications.

8.1 RECORDS AND BLOCKS

The electromechanical principles of magnetic tape units are well known, and there is nothing to be gained from restating them here. We are, however, concerned with the layout of magnetic tape; thus, it is helpful to appreciate the reason for blocked data. Magnetic tape can be read or written only while it is moving, during which time the data is transferred in input or output mode. Unfortunately, this data cannot be processed simultaneously with its transfer and so a limit is imposed on the amount that is contained within one transfer. This is a "block" of data, and is a physical concept in that it is transferred automatically by the computer. At the end of a block the computer ceases to transfer data because an "end of block marker" is detected.

The size of the block within a file is decided at the time the file is created, and thereafter remains constant. The main deciding factor is the amount of core storage that can be spared for holding the block (the buffer area). Several buffer areas may be needed if a number of magnetic tape units are employed simultaneously.

Similarly, two buffer areas are needed if double-buffering is used. Double-buffering means that one buffer area is being loaded or unloaded by the tape at the same time as the other buffer area is being processed. This technique obviously speeds things up, since the time between blocks is reduced.

After the cessation of data transfer at the end of a block, the tape continues to move so that the read/write head becomes adjacent to the "interblock gap" in which no data is recorded. The size of the interblock gap is decided initially by the rate at which blocks can be made ready in core storage for transfer to tape. It is important that the block-to-gap length ratio is as high as possible, since this affects the time taken to process the tape. The interblock gap has a maximum length so that in many cases this length is the deciding factor. A very considerable difference in read/write time is experienced with variation in block length; for example, with a given tape, it is quite possible for a change in block length from 100 to 1000 characters to reduce the reading time from 9.4 to 2.4 minutes for the same number of records. Actual block lengths used vary from one computer and from one application to another, but in round terms they lie between 100 and 2000 characters.

The magnetic tape record is a logical concept, its size being related solely to its application and in no way controlled by the hardware. Thus, a record may be smaller or greater in length than a block, although in practice it is usually smaller. When a record turns out to be larger than the block, it is advantageous to form split records (Section 6.2). The relationship of blocks to records is completely flexible and, even within the one file, is not necessarily constant. As explained earlier (Section 6.2), record lengths may be variable, owing to their holding either (1) a variable number of fields, or (2) fields of variable length, or (3) both variable number and length. In the extreme case where fields, records, and blocks are all variable, it is necessary to employ various "markers" so as to indicate their end points.

To summarize, a data block may consist of:

1. A fixed number of fixed-length records (fixed–fixed block).
2. A fixed number of variable-length records (fixed–variable block).
3. A variable number of fixed-length records (variable–fixed block).
4. A variable number of variable-length records (variable–variable block).

Examples of these alternatives are shown in Fig. 8.1. Whenever possible, the length of a variable block should be close to the maximum length so as not to waste core storage space and time during the processing runs.

8.2 FILE UPDATING AND AMENDMENT

The same principles of updating and amendment apply to magnetic tape files as were described for direct access files in Section 6.3. Similarly, magnetic tape files can be classified according to their contents in the same way as direct access files; i.e., into movement, transition, and master files.

File-Reel Relationship

The fact that magnetic tape is in limited lengths (i.e., reels) is purely for the convenience of handling it. Reels are up to 3600 feet in length, and this length of tape (0.0015-inch thickness) can be accommodated on one 10.5-inch diameter spool. Makes of thicker (0.002-inch) tape are in lengths of up to 2400 feet per spool.

There is no fixed relationship between a reel and a file. It is quite normal for long files to occupy several reels, called "multireel files." On the other hand, it is often convenient to hold several smaller files on one reel, called

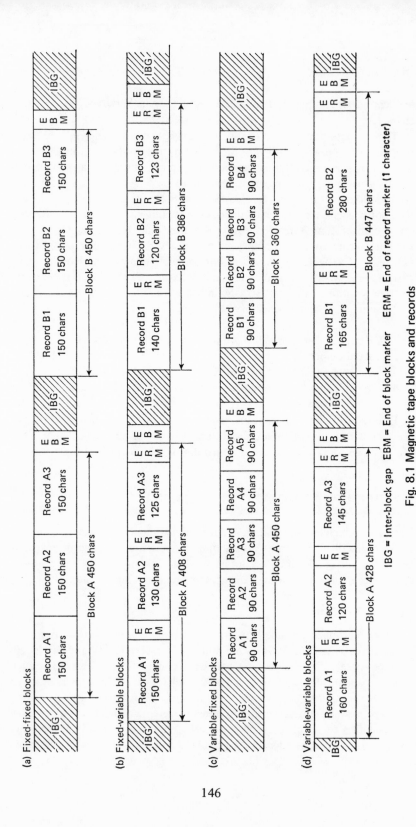

Fig. 8.1 Magnetic tape blocks and records

IBG = Inter-block gap   EBM = End of block marker   ERM = End of record marker (1 character)

146

"multifile reels." These arrangements cause no difficulty, provided the reels are properly organized from both the human-handling aspect and the magnetic labeling by the software. Nevertheless, it is often convenient to have only one file per reel, if necessary using shorter reels so as not to waste tape. This arrangement is unavoidable when files have to be processed against each other in the same computer run; i.e., the reels have to be loaded onto separate tape drives.

Updating and Amendment Runs

The only viable method for comparing records on two or more magnetic tape files is by first sorting them into the same sequence. Thus, when using a movement file to update or amend a master file, it must be sorted into the same sequence as the master file. For some computer runs, it is also necessary to sort the movements into another sequence within the primary sequence. This applies, for example, to stock updating and allocation; in this application some movements have priority, and therefore the sort must be into date sequence within stock number sequence.

Brought-Forward–Carried-Forward Files Method. Because it is impractical to rewrite master file records on the same tape reel as they have just been read from, it is necessary to employ two tape drives for handling master file data during an updating or amendment run; this is shown in Fig. 8.2. The brought-forward master file is read along with the movements file, both in the same sequence. When record keys coincide, the master record is updated or amended by the related movement(s), and the new version of the record is written onto the carried-forward file. Master records that are not changed are copied identically onto the carried-forward file. The carried-forward file from one run becomes the brought-forward file for the next run. When using this method for updating and amending a master file, it is necessary to make special arrangements for dealing with movements that do not have a corresponding master record. This condition often indicates that an error has been introduced at some point in the system.

Changes Tape Method. This is an alternative to the preceding method, using four tape drives instead of three. The principle employed is to read the movements and the master file as before, but to write only changed master records to a "changes" tape. During the next updating run, the "changes" tape, the master file, and the next lot of movements are all read. Any records from either the changes tape or the master file that are changed are written onto another changes tape. Thus, after the second week's run, the latest changes tape holds an up-to-date version of every master file record that has been changed during week 1 and/or week 2.

This procedure is shown in Fig. 8.3 for an application in which the movements are read weekly but the master file is updated only every four weeks.

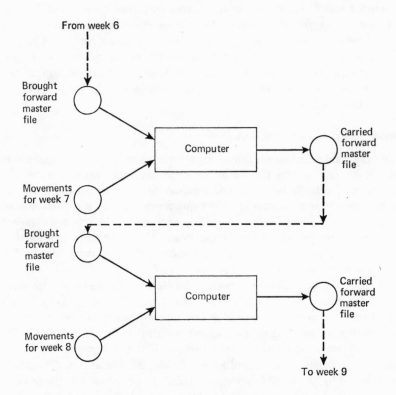

Fig. 8.2 Brought-forward and carried-forward master files

Provided it is not necessary to have the master file in an up-to-date condition except at the four-week period end, the method can be used as a means of reducing the overall processing time. This is especially true if the movements tend to apply to only a small proportion of the master records (low activity file). Often this method cannot be used for amending a master file because normally the amendments must usually take effect immediately.

As a rough guide to comparative times, if the volume of movements in each run is about one-tenth of the master records, and the same keys predominate on the changes tape, the changes tape method would take approximately three-quarters of the time of the brought-forward–carried-forward method. An example of the various file contents of a stock updating run using the changes tape method is shown in Fig. 8.4.

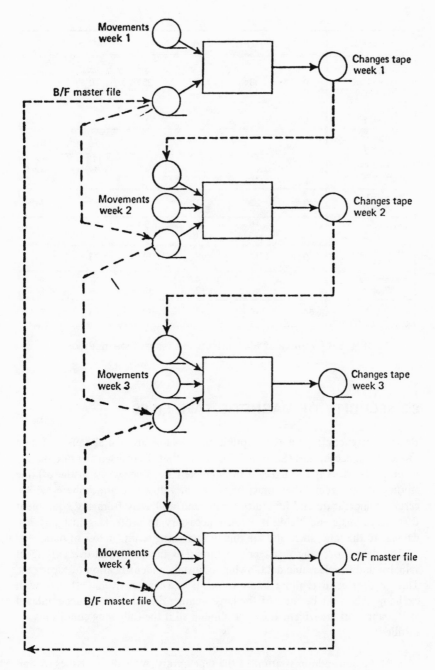

Fig. 8.3 Changes tape method

149

Stock No.	A46	A47	A51	A52	A53	A61
B/F Master from last month	50	38	10	25	70	94
Movements Week 1	+10 +6	—	+ 3	—	—	-8 -2
Changes Tape Week 1	66	—	13	—	—	84
Movements Week 2	-2 -11 -4	+1 +4 -10	—	—	+10 -17	+2 -12
Changes Tape Week 2	49	33	13	—	63	74
Movements Week 3	—	+9 -23	-4 +16	-5 -7	—	—
Changes Tape Week 3	49	19	25	13	63	74
Movements Week 4	-1 +16	—	+4 +3	+2 +3	—	-2 +5
C/F Master to next month	64	19	32	18	63	77

Fig. 8.4 Example of file contents in changes tape method

8.3 SECURITY OF MAGNETIC TAPE FILES

Reels of magnetic tape are susceptible to the same accidental causes of data loss as direct access files (Section 7.7), except that the overwriting of data due to program errors is less likely. Considerable care is necessary in the off-line storing of tape reels; they must be stored in a dust-free atmosphere, within certain temperature and humidity limits, and well away from any equipment that has a magnetic field. It is also necessary to protect against physical damage to the tape, since any kinking or stretching results in loss of data.

The data processing manager's nightmare is the accidental overwriting of valuable and irreplaceable data, owing to mistaken identification of tape reels. This type of error is more likely to occur in a tape environment than with exchangeable disks because of the large number of reels that are accumulated as the applications widen. It can be eliminated if the following conditions are fulfilled:

1. Meticulous administration in the tape library, with all reel issues to, and

receipts from, the computer operators being controlled by the tape librarian.

2. Correct utilization of magnetically recorded tape labels as specified by the manufacturer and used in his software.

3. The use of "file protect rings" on the reels of tape that are to be written during a run. These rings are attachable to the center part of the spool and are not removed unless a reel is to be written on.

4. The employment of the "father–son" technique when updating or amending master files. Although this technique does not in itself prevent overwriting, it is a valuable security measure against this and other types of accidents.

Grandfather-Father-Son Technique

The principle behind this technique is simply to retain a brought-forward tape for some time after the creation of the corresponding carried-forward tape. It is customary always to have *at least* two "generations" of tape in existence at any time, but three generations give added protection. Thus, during an updating run, there are three tapes in temporary existence:

1. The "son" tape in the process of being created on a fresh reel; i.e., the carried-forward tape.

2. The "father" tape, which was the previous week's son tape, being read; i.e., the brought-forward tape.

3. The "grandfather" tape, which was the previous week's father tape, retained on the shelf for security purposes.

When a son tape has been fully created and validated, the grandfather tape becomes eligible for scratching; i.e., reuse for other data. When the next updating run starts, the son becomes the father and the father becomes the grandfather tape. It is necessary to retain also the movement tape from the previous run. By adopting these measures, it is possible to recreate a son tape by rerunning the previous run's movement tape and father tape. This enhances the security of the system by safeguarding against loss of data through tape damage or accidental overwriting. A higher degree of security is achieved by retaining the tapes for more generations, although practical considerations of tape cost and shelf-storage space must be considered.

Dumping

An important consideration in long updating runs is the possibility of a mishap occurring toward the end of the run, necessitating a restart from the beginning. This can be very time wasting and frustrating, and so if a run takes more than a half hour, it is usually worth splitting it into sections. This means that at the end of each section, the contents of the core storage are "dumped" onto a reel of tape; in other words, exact copies are made of the program, intermediate results, and any other contents of core storage. Sections may occur naturally as a result of the master file's contents, and it is often convenient to use these as the dump-point sections, although there are no reasons why any other points should not be used.

Dumping is made onto a spare tape, or if a spare tape drive is not available during the run, it can be made onto the carried-forward master tape. The restart procedure involves backspacing the master file tape to the dumped data (if present on this tape), or to the start of the section during which the trouble occurred. The core storage is then reloaded with the program and other data in the exact form that they were in at that point of the run, the brought-forward tape and the movement tape are backspaced to the start of the section, and the run recommences. Dumping and restart procedures are well covered by manufacturers' packages and operating systems.

8.4 MAGNETIC TAPE SORTING

When using magnetic tape systems, sorting is a more intrinsic feature than with direct access systems. The serial nature of tape files makes it axiomatic that their processing be in sequence, with the result that the processing runs are interspersed by sorting and merging runs. The principle reasons for magnetic tape sorting are the following:

1. To arrange a movement file in the same sequence as a master file before an updating or amendment run. This is unavoidable unless the volume of movements or the size of the master file enables either of them to be held completely in core store throughout the run; this situation is highly unlikely. Examples of this are:
   (a) Weekly earnings updating a payroll file.
   (b) Invoice details updating an accounts receivable file.
   (c) Stock issues and receipts updating a stock-level file.
   (d) Sales tax changes amending a commodity master file.
   (e) Job tickets updating a work-in-progress file.
2. To list or tabulate a master file in a different sequence from that in which it is usually held. This happens, for instance, with a stock file that is normally held in part number sequence but is needed in bin number sequence for stock-taking purposes.
3. To collate items of a similar nature for the purpose of comparison or summarizing. For example, sales records are sorted into area within class-of-

customer sequence so as to facilitate comparison of sales to similar customers in the same locality.

4. To merge two or more files in order to create one larger file. This occurs quite frequently during master file creation when the data comes from a number of different sets of source documents.

When creating a commodity master file for subsequent billing purposes, for instance, it might be necessary to merge a price file, a sales tax file, and a discount structure file. The reason is because these sets of data originated from different sources and hence were initially put onto separate files.

5. To highlight certain factors by bringing them to the beginning of a file. This reason for sorting differs from the others in that the sorting key could well be a quantity or value rather than an indicative field. An instance of this is the sorting of job costs in reverse sequence of their values, thereby putting the higher costs at the top of the job-costing list.

Sort Generator Programs

The programs required for carrying out magnetic tape sorting may be written in-house, but are also supplied to the user as part of the software supporting the hardware. They are usually in the form of a "sort generator" program that can create the most suitable type of sorting routine for the job in hand. The "sort generator" functions by being provided with a set of parameters by the user when he wishes to set up a sorting run. The parameters cover the details of the data to be sorted and the hardware to be used, and include:

The number of records to be sorted.
The sizes and positions within the records of the fields comprising the sorting key.
The degree of initial sequentiality (if any).
Forward or reverse sorting indication.
The number of tape drives available.
The amount of core storage available.

Tape-Sorting Times

An important factor entering into the design of data processing systems based on magnetic tape is the tape-sorting time. This time cannot, of course, be predetermined precisely because it does depend to some extent upon the randomness of a particular set of records at the time of sorting. The time norm, however, is available from computer manufacturers' manuals in an exact enough form. The factors that affect sorting time are:

The number of records to be sorted.

The average size of the records.

The number of tape drives available.

The transfer rate (read/write speed) of the tape and drives.

The amount of core storage available.

The block size and blocking method.

8.5 EXERCISES

Problem 1. Magnetic Tape File Sizes and Times

A file consists of 50,000 records of 200 bytes and is held on magnetic tape in blocks of B bytes with an interblock gap of 0.6 inch. For each of the tape drives specified below, calculate (i) the amount of tape required, and (ii) the time taken to read the file into a computer, assuming that a single buffer area is available equal to the block size and that there is no delay between blocks. Ignore the need for markers.

PEAK TRANSFER RATE, BYTES PER SECOND	BYTES PER INCH	BLOCK LENGTH (B), BYTES
(a) 60,000	1600	2000
(b) 15,000	800	1000
(c) 180,000	1600	5000
(d) 30,000	1600	4000

Solution to Problem 1. From the figures given, it is possible to calculate further quantitative characteristics of the tape drives:

$$\text{Tape speed, inches/second} = \frac{\text{bytes/second}}{\text{bytes/inch}}$$

Interblock gap time = 0.6 ÷ tape speed

Records per block = B ÷ 200

Time per block = B ÷ bytes/second

Number of blocks = 50,000 ÷ records/block

Block length = B ÷ bytes/inch

Thus, the amount of tape (i) = number of blocks (block length + 0.6), and time taken (ii) = number of blocks (time per block + interblock gap time). Tabulating these figures, we have

Tape Speed, In./Sec.	IBG Time, MS	Records Per Block	Time Per Block, MS	No. of Blocks	Block Length, Inch	Amount of Tape, Feet (i)	Time Taken Min. Sec. (ii)
(a) 37.5	16	10	33	5,000	1.25	771	4.5
(b) 18.75	32	5	67	10,000	1.25	1542	16.30
(c) 112.5	5.3	25	28	2,000	3.125	621	1.6
(d) 18.75	30	20	133	2,500	2.5	646	6.52

Problem 2. Records and Blocks

The details of the states given in Appendix I are to be written onto magnetic tape. Work out the length of the records and blocks required on the assumption that:

(A) There is one record per state.
(B) Each letter occupies one byte, and each digit occupies half of one byte, but bytes cannot be shared between two fields.
(C) Fixed-length and variable-length records are to be fitted into fixed and variable blocks of between 300 and 400 bytes (as per Fig. 8.1).
(D) The produce consists of separate fields, one field for each produce.

Draw a diagram to show the details of VIRGINIA'S record as accommodated in (i) a fixed-length record that can hold the longest fields of all states, and (ii) a variable-length record of minimum possible length, including all markers.

Solution to Problem 2. The first step in solving this problem is to analyze the data so as to ascertain the maximum field sizes and their pictures.

FIELD	PICTURE	MAXIMUM	ACTUAL MAXIMUM
State	A(20)	A(20)	(DISTRICT OF COLUMBIA)
Area	9(6)	571,065	(ALASKA)
Population	9(5)	19,953	(CALIFORNIA)
Capital	A(14)	A(14)	(JEFFERSON CITY and SALT LAKE CITY)
Produce (1–6 fields)	A(13)	A(13)	(MANUFACTURING)

Fixed-fixed block: The size of the fixed-length record, as determined by the pictures, is $20 + 3 + 3 + 14 + (13 \times 6) = 118$ bytes. A convenient block size would be 354 bytes; i.e., one containing three records, since this lies within the limits mentioned in (C).

Fixed-variable block: The mean size of these variable-length records is about 50 bytes. We could therefore consider having seven variable-length records composing each block. The first seven states in the list, for instance, would occupy around 350 bytes (including "end of record markers").

Variable-fixed block: This arrangement would consist simply of having one or more fixed-length records of 118 bytes each per block; thus the blocks would be 118, 236, 354 bytes, and so on. In this particular example, variable-fixed blocks are not of any obvious advantage.

Variable-variable block: The principle behind this arrangement is to decide upon a maximum block size, say, 400 bytes, and to fill the blocks with variable-length records until no more can be accommodated. This results in variable-length blocks of roughly 350 through 400 bytes. The first block, for instance, would comprise the first nine records and occupy 393 bytes.

VIRGINIA'S record: See Fig. 8.5.

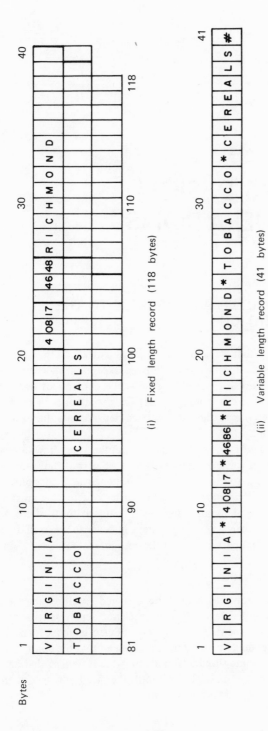

Bytes

(i) Fixed length record (118 bytes)

(ii) Variable length record (41 bytes)

Fig. 8.5 Examples from Problem 2

157

CHAPTER 9

SYSTEMS DESIGN CONSIDERATIONS

9.1 THE COMPUTER AS A SERVICE TO THE ORGANIZATION

Along with most other departments in an organization, the data processing department is not an end in itself. With the exception of when, if ever, it is doing chargeable work for outside agencies, the department does not appear to be directly involved in making money—nor sometimes in saving money— and its role as a service to the other service departments does not give it the most exciting of images.

The initial impact of a newly installed computer often draws a considerable amount of attention toward a new system. It is part of the systems analyst's function to insure that this interest is maintained at a sensible level there-after; the computer's opening role as the center of attention must not be fol-lowed by a lifetime of remote isolation. What can the systems analyst do when designing the data processing system in order to make all departmental staff feel that they are part owners of the computer?

Encourage Communication

Whenever suitable, routines should be designed to allow amendments and inquiries to be accepted regularly from the user departments. This does not imply that the departments should be encouraged to request special amend-ments in a never-ending stream, but it does mean that the ability to deal with preplanned amendments should be built into the system so that amendments can be dealt with automatically without the need for special unforeseen programs and unexpected computer runs.

The ability to deal with amendments in a smooth and effortless manner is based upon the design of open-ended routines. It is, of course, virtually impossible to deal with every conceivable amendment in this way, but nevertheless the majority can be handled. Examples of amendments that could well be expected to occur in data processing systems are:

1. The insertion and deletion of items held on file.
2. Changes in the percentages and structure of sales tax.
3. The insertion of additional fields into existing records; for example, telephone numbers inserted into name and address records.
4. Considerable variation in the content and layout of management reports.
5. Changes to individual records, perhaps involving only one particular field in a record; for example, the correction of an erroneous street number in an address file or of a date of birth in a personnel file.

In a similar way, the need for genuine inquiries must be ascertained, and methods for coping with them must be included in the system. Every effort should be made to dispel the aura of remoteness that can quickly become associated with data processing departments. Although it is extremely difficult to envisage all theoretically possible inquiries, a large proportion should be allowed for in the system. If necessary, a generalized information retrieval system such as IBM's GIS can be employed. By finding the characteristics of such a system beforehand, the systems analyst is able to design file contents and layouts to suit it.

Users must not be given the impression that data processing is a rigid and inflexible methodology. Instead they should be assured that data processing systems can respond to their needs in a rapid and intelligent manner. By encouraging communication between users and the data processing department, it is easier to visualize impending changes and special requirements in good time, thereby avoiding the last minute frantic programming often associated with such contingencies.

In a large or dispersed organization it is beneficial to maintain a regular and formal communication link such as user department meetings with the data processing staff, say, bimonthly. On some occasions there will be little to discuss; on others, very significant new information and worthwhile ideas will emerge.

Encourage Participation

In connection with inquiries and other requirements, arrangements should be made for the user department staffs to obtain their needs from the computer automatically and without making special requests. This can be done in two main ways.

1. By the submission of standard inquiry forms to the data processing department. These inquiries would be dealt with as a matter of routine and at predetermined times, most probably when the relevant file was already available to the computer.

2. By the writing of programs by the user department staffs, probably in COBOL, FORTRAN, or other high-level language. This arrangement is particularly efficacious for meeting the needs of technical staffs such as design engineers and research workers. In cases where the computer is to be used for relatively small amounts of technical work, it is easier for the technical man to learn to program in, say, FORTRAN than for the systems analyst to learn the technical work.

Another aspect of participation in data processing is the working together of the user and the data processing staff in project teams. This was mentioned earlier in connection with systems investigation, and the close relationship built up at this time should be maintained into the systems design and implementation stages. Although the user staff cannot be expected to play much part in designing data processing systems, its members can be useful in other ways, such as checking the exactness of proposed output and in preparing test data and file data.

Arrange "On Request" Output

As applied to data processing, the principle of management by exception has many advantages and should therefore be implemented whenever possible. There are, however, some situations in which a large volume of printed output is unavoidable. This generally occurs where there is a need either for readily available reference information in the form of a directory or for special reports in order to analyze mentally the information therein. A danger here is that reference information may swamp the recipient, and as a consequence, the principle is seldom applied. Similarly, special "once only" reports become perpetuated, resulting in an ever-increasing amount of printed matter being prepared by the computer.

A solution is to make these types of computer outputs into "on request" items. The applicants for special reports must then submit a request on a standard slip. As a reminder of this service, and as a guide to the latest dates in the week, month, etc., by which requests should be submitted, other regular computer output may have appended details printed by the computer.

If the special reports occupy very little computer processing time, it may be worth preparing them automatically, but they should be distributed only to persons making specific requests on that occasion.

9.2 INTERRELATIONSHIP OF FILES

It was suggested in an earlier chapter that the files are the skeleton of a data processing system. Just as a skeleton has to fit together in a flexible manner, so must files, in order that they can perform their function. This concept applies to all classifications of files (Chapter 6), and most of all to master files because these are the backbone of the system.

A paradoxical feature of systems design is that a file layout cannot be decided exactly until all the related applications have been designed, and yet no application is fully designed until its files are laid out precisely. This problem is overcome by iterative methods of design by which the file contents and layouts are decided gradually, fields being inserted as each application is designed.

When designing the file layouts and contents, the systems analyst has to bear in mind the following points:

1. The contents of a file must.be adequate to meet the requirements of all routines that will use it; these may include routines that will be implemented at a later date. A parts master file, for example, may initially be employed only for determining the raw material requirements of a production program. If it is known that machine loading will shortly be implemented as a data processing routine, it could well be worthwhile to incorporate the part's machining operation times on the file from the outset.

2. If there is doubt as to the certainty of a field's being required in the records in a file, it is usually simpler to include it when the file is created than to have to insert it later. A possible disadvantage in this is the perpetual updating or amendment of an unused field. An alternative is to leave a space for the field, and although this is wasteful of storage, it is generally preferable to having to reorganize the whole file later.

An example of this latter situation would be a stock file in which a blank field is left for the reorder level of each item because these figures are unknown when the file is created.

3. Separate files that are to be sorted together must have their keys in the same relative position within the records. This ruling also applies sometimes to different files that are to be matched or merged by the use of the computer manufacturer's standard software. The precise requirements as regards this point should be ascertained from the manufacturer.

4. The magnetic tape block length or the bucket size of a direct access storage device should be planned so as to be suitable for all computer runs that use the file. This factor hinges on the core storage taken by the most complex programs in the system. The analyst cannot be expected to know the exact program sizes at the systems design stage, but approximations can be made so that block or bucket sizes can be decided.

5. Upon occasion, files contain records or fields that are not actually used in the regular data processing routines. These insertions are often for reference purposes in connection with inquiry and interrogation procedures. In other cases their presence is as a security precaution, such as when a range of items has recently been recoded and, as a consequence, it is vitally necessary not to confuse the identities of the items.

File Record Usage

As a means of deciding the format and insuring the completeness of file records, it is helpful to fill in a "file specification sheet" for each file that is to be created (Fig. 9.1). This document shows the contents of the file's records and its utilization by the processing routines. Each field is specified and cross-referenced to the file utilization, in which it is ticked off against the routines that make use of it. In this way the chance of a field's being accidentally omitted from a file is minimized. In systems where the routines consist of a large number or a complex arrangement of computer runs, the fields could be ticked off against the runs instead of the routines. The precise method of specifying the records depends upon the storage device employed and the characteristics of the computer's addressing system; Fig. 9.1, however, gives a general indication of the basic requirements.

9.3 INTEGRATED DATA PROCESSING

The concept of integrated data processing (IDP) has been with us ever since the first business applications were applied to computers. It has, however, in many cases, proved easier to conceive than to translate into a working system. There are nevertheless an increasing number of companies in which this concept is achieving reality, and also a growing awareness that this is very often the best basis on which to build a data processing system. What, then, is meant by integration within the framework of data processing?

It is often believed by people in the computing world that integration means "doing everything by computer"—forcing every application into the "total system." The nebulous assumption supporting this idea is that the more applications are "computerized," the more profitable they become,

FILE SPECIFICATION				
File name	Ref.	Classifications (delete as necessary)	Normal sequence	Created by run ref.
Product master Cost File	T2	Movement Serial Transition Sequential Master Random	Product code	A3.

Field ref.	Field name	Picture	Byte positions	Remarks
1	Product code	99999	1-3	
2	Std. labor price $	99.999	4-6	To nearest 1/10 cent
3	Std. material price $	99.999	7-9	To nearest 1/10 cent
4	Max. discount %	99.999	10-12	Allows for fractions
5	Selling price	999.99	13-15	$ and cents
6	Description	X(20)	16-35	
7	Manufacturing group	9	36	Eight groups
8	Batch size	999	37-38	
9	Factory no.	9	39	Four factories
10				
11				
12				

FILE UTILIZATION													
Routine code	Routine name	Fields used – refs.											
		1	2	3	4	5	6	7	8	9	10	11	12
K	Sales analysis	✓	✓	✓	✓	✓		✓					
T	Costing	✓	✓	✓		✓		✓	✓				
H	Invoicing	✓				✓	✓						
E	Order handling	✓				✓	✓	✓	✓	✓			
G	Production planning	✓						✓	✓	✓			
M	Production control	✓						✓	✓	✓			
P	Sales forecasting	✓	✓	✓	✓	✓	✓						

Fig. 9.1 File specification and utilization

since the computer is available at no extra cost. This assumption is occasionally correct, but the systems analyst must, if necessary, disabuse himself of the belief that it is invariably so.

Instead of regarding integration as the total system, it is wiser to aim at a system that unites all the applications already included, and one that is also as "open ended" as possible. The latter characteristic implies that further applications can be built into the system with a minimum of difficulty. The main aim of integration should be the uniting of existing data processing applications; attempting to encompass every possible application is not necessarily desirable.

Provided our IDP system is open ended, it is not necessary to implement all applications within it simultaneously. These may be transferred onto the system as and when it is convenient to do so, but at each stage an allowance should be made for those that follow. The implementation of "closed" or isolated applications without regard for their successors will undoubtedly cause problems later in interfacing the old and new routines. For instance, the implementation of stock control with no provision for recording a stock usage history will undoubtedly hinder the later implementation of an automatic reordering system.

We must also guard against assuming that integration necessarily entails the employment of a large computer or, for that matter, of a computer at all. In principle, a manual system can be fully integrated; in practice, this is difficult because of the slowness of manual operations and the problems of high-speed communication between departments. The size of the computer required depends on the volume of its work and files; the degree of integration achieved depends upon the skill of the systems analyst.

Design Features of Integration

When designing an integrated system, the analyst is well advised to bear in mind certain basic principles, as outlined below.

Source Data. This enters an IDP system once only, and thereafter is processed along with other data in order to produce the required output results. It is very inefficient to input any significant amount of source data more than once because this procedure occupies a relatively large amount of computer time. Typically, the data from a production job ticket should enter the system once only and then be processed as often as necessary to provide information for job costing, payroll, work-in-progress control, and so on. It is often the case that the source data is written onto magnetic tape and then read by the routines that use it. Since magnetic tape can be read very much faster than source data media, this principle is acceptable as good data processing practice.

Phasing of Routines. Each routine in the IDP system has a phase relationship with other routines—there are obvious cases where one routine simply cannot function until another has reached a certain stage. In other cases the efficiency of a routine is jeopardized if its related routines are out of phase with it. An example of this is when a stock file fails to be updated with receipts into stock before allocations are made from it. The consequence of this situation is that items are incorrectly designated as out of stock, and sales orders are lost or the dispatch of goods is delayed.

Interrelated Files. These were described in Section 9.2.

Common Coding. Because integration entails the connecting together of the various routines and applications through their common data base, it is important that the identical code number be used by all routines for a given item. In cases where practical considerations prevent this, the system must be capable of accepting any one of the alternative code numbers as the key to the identity of the relevant item (Section 7.5).

If a changeover from one set of code numbers to another is made, care must be taken to insure that the data processing system is capable of accepting either of the code numbers applicable to an item for some considerable time after the changeover date.

Output Reports. The printing of common information by the computer for use by various departments should be coordinated. Owing to the computer's ability to process data quickly, output reports are more up to date than from other systems. This may result in what appear to be contradictions. Therefore, as a safeguard against unjustified criticism, the information should be either time-dated or produced at one and the same time in the form of a combined report. For instance, a product stock-in-hand report for the sales department must concur with that prepared for the purchase department.

Flexibility. An IDP system must be not only capable of expansion in terms of the data volumes but should also be capable of handling unforeseen eventualities as far as is reasonably possible. This need for flexibility is pertinent to any system, even if it comprises only an isolated application. It is markedly more true of an integrated system due to the danger of creating an overrigid, self-contained system that cannot be employed for special purposes except with excessive difficulty. An example of this is the inaccessibility of file data to anything except the computer. Obviously, this is a technical truism, but the point being made is that all computer file records should be readily available for manual inspection—perhaps for completely unanticipatable reasons.

The extent to which most presently existing data processing systems cannot cope quickly enough with a change in circumstances is amazing. A postal strike, for instance, results in a flood of telephone inquiries to insurance companies from their customers regarding their automobile insurance coverage. The required information is then found to have been locked away on a computer file, and is therefore completely inaccessible manually. The result of this situation is widespread confusion about the customers' legal position in the event that the mails have delayed delivery of their policies.

This comes back to the systems analyst and the need for him to envisage all eventualities that are reasonably possible. Experience of previous unusual circumstances is invaluable in this respect, and it is partly the users' responsibility to suggest fail-safe methods for their routines in the event that difficulties arise.

9.4 SYSTEM DEFINITION AND DOCUMENTATION

The system definition embraces the formal detailed description of the data processing system in the form of written documents. It is essential that the system be documented, for three main reasons:

1. As a basis for obtaining official top management approval of the system prior to its implementation.
2. As a means of disseminating information about the system to persons who are involved in its implementation and operation.
3. As a reference document for the future, bearing in mind that its readers may be staff who are completely new to both the system and the organization.

The system definition documents are, for the most part, prepared during or after the design of the system. It is, however, beneficial to have the proposed contents in mind right from the commencement of design, as this will guide the systems analyst in his deliberations. When preparing the definition, its layout should be arranged to suit the three purposes listed above. The information provided for top management obviously need not be as detailed as that for the programmers; the emphasis in one case is on aims and advantages, and in the other is on precise computer run specifications. It is, in any case, unlikely that the run specifications will be ready by the date at which the system is presented for top management approval (Section 12.4). Similarly, the information needed by the staff (see item 2 above) is not always the same. Included in this group are not only the data processing staff but also other departmental staff.

The preparation of the system definition inevitably takes a fair amount of time, so the systems analyst should avoid duplication of effort. Each section of the definition should be written with all its readers in mind so that there is no rewriting of the same information. By carefully sectionalizing the document, the appropriate sections can be collated for presentation to the particular recipient, omitting the sections that are of no interest to him. No hard-and-fast rules can be laid down regarding the distribution of sections; the systems analyst must judge each section's pertinence in relation to its possible recipients.

Amendments to the System Definition. It must be remembered that the system definition is likely to incur amendments, particularly in the more detailed sections. When these occur, a memorandum should be sent to the appropriate staff in order to explain them, and if necessary a whole section should be redistributed to replace the original. To facilitate the distribution of amendments and new sections, a list of recipients can be included at the start of each section. The problem of amendment distribution is, of course, another good reason for restricting the distribution of sections to interested parties only.

Sectionalized Documentation of the System Definition

It is not the intention to describe here each section of a system definition; this is done in other chapters. A list of suggested sections together with a reference to the relevant section of this book follows below:

1. List of contents and amendments record.
2. Distribution list of each section of the system definition.
3. Aims and advantages of the data processing system (Sections 3.1 and 12.2).
4. A general description of the overall system, avoiding technical jargon but emphasizing departmental staff participation and system security (Sections 7.7, 8.3, and 9.1).
5. The specifications of all equipment involved, and a brief explanatory note of new or unusual items of hardware (Section 11.5).
6. A summary of the estimated costs of the old and new systems (Section 12.1).
7. An implementation timetable (Section 13.6).
8. A schedule of the computer routines (Section 11.3).
9. A flowchart and a description of each computer routine (Section 10.5).
10. Instructions to the user departments (Section 13.4).
11. Punching instructions (Section 10.3).
12. File specifications (Chapter 6 and Section 13.2).
13. Distribution list of computer output (Section 13.4).
14. Run specifications (Section 10.6), each run being separable for distribution to the appropriate programmer only.
15. A report on the future outlook, and possible enhancements to the system in relation to the organization's expansion and development.

9.5 FINANCIAL ASPECTS OF DATA PROCESSING

It is only too easy for computer people to neglect the financial considerations of data processing in their pursuit of technical excellence. Data processing systems should be subjected to the same degree of cost justification as any other project. In view of the distinct possibility that their costs will be very high, it could be argued that they should be given even more scrutiny than most projects.

The esoteric nature of data processing and its aura of sophistication, even though completely unjustified in many cases, has engendered a reluctance to consider the question, "Is a data processing system a sound financial proposition for us?" At a later stage the question becomes, "Is our data processing system producing the benefits that we intended?" These two questions indicate the main stages at which financial and benefit appraisals should occur—first, before the decision is taken to install a data processing system, and second, after installation in order to insure that the intended benefits have really been acquired without at the same time incurring extra unforeseen costs.

When these factors are being considered, it is important for the systems analyst to work closely with the company accountant,* since these considerations are closely connected with financial accounting.

Discounted Cash Flow (DCF)

In order to make a quantitative comparison of two or more projects, it is necessary to adopt a scientific method of appraisal. This is, of course, primarily the responsibility of the accountant, but it is advantageous to the systems analyst to have an understanding of the principles involved. The techniques are embodied in a number of computer packages intended to remove the burden of extensive calculations from the accountant's shoulders. Examples of these packages are Honeywell's DISCET and RECAP, and ICL's PROP.

The DCF techniques,* although providing a rational means of appraisal, are not infallible and do not really remove the onus of financial decision making from management. They do, however, provide figures upon which decisions can be based while taking other factors into consideration. The technique described below is the Net Present Value (NPV)* method, which is based on the future cash flows associated with a management decision to proceed with a project.

*Clifton and Lucey, *Accounting and Computer Systems,* Petrocelli Books, New York, 1974, chap. 8.

The principle behind NPV is that money received now is worth more than money received in the future. Consequently, all future receipts and expenditure, no matter when they will occur, must be converted to their present-day value. To do this, it is necessary to reduce (discount) future sums of money by a certain rate of discount (discount rate). For instance, the sum of $3000 received two years hence is at present worth $2479 if the discount rate is 10 percent; i.e., $2479 \times 1.1^2 = 3000$.

If the preceding calculation is applied to all cash flows (+ and − corresponding to receipts and expenditure) and the present values summed, this total is the NPV of the project. A positive value for NPV indicates a capital gain for the project, which—ignoring risk and uncertainty—means that the project is financially worthwhile if the existing cost of capital is 10 percent. The foregoing explanation is a very much simplified version of proper investment appraisal, but nevertheless it gives the systems analyst some ideas about what is involved.

9.6 EXERCISES

Problem 1. File Specification and Utilization

(A) Enter a file specification sheet (as in Fig. 9.1) to cover a materials purchasing file (ref. E3) to be held in material code sequence. This file is created by run reference D5, and has the following contents:

Material code of three letters followed by four digits, from one through four suppliers, each having a four-digit account number; standard cost price per unit in dollars and cents up to a maximum of $7.00; the current discounts from each supplier (up to 20 percent in tenths); the material category (one of six); and the description of the material (this can be as long as 40 characters).

(B) Enter a file utilization chart for the file in (A) on the assumption that it is utilized by the following routines:

B (purchase control): all fields used.

G (production planning): material code, description, and category.

M (purchase accounting): all fields except category.

P (stock evaluation): material code, cost, and category.

Solution to Problem 1. See Fig. 9.2.

FILE SPECIFICATION

File name	Ref.	Classifications (delete as necessary)		Normal sequence	Created by run ref.
Materials purchasing file	E3	Movement ~~Serial~~ ~~Transition~~ Sequential Master ~~Random~~		Material code	D5

Field ref.	Field name	Picture	Character positions	Remarks
1	Material code	AAA9999	1-7	
2	Description	A(40)	8-47	
3	Category	9	48	Six Categories
4	Std. cost price	9.99	49-50	
5	Supplier A/C no.	9999	51-52	Up to four suppliers
6	" disct.	99.9	53-54	Up to 20%
7	" A/C no.	9999	55-56	
8	" disct.	99.9	57-58	
9	" A/C no.	9999	59-60	
10	" disct.	99.9	61-62	
11	" A/C no.	9999	63-64	
12	" disct.	99.9	65-66	

FILE UTILIZATION

Routine code	Routine name	Fields used — refs											
		1	2	3	4	5	6	7	8	9	10	11	12
B	Purchase control	✓	✓	✓	✓	✓	✓	?	?	?	?	?	?
G	Production planning	✓	✓	✓									
M	Purchase accounting	✓	✓		✓	✓	✓	?	?	?	?	?	?
P	Stock evaluation	✓		✓	✓								

Fig. 9.2   Solution to Problem 1

170

CHAPTER 10

DESIGN OF DATA PROCESSING SYSTEMS—1

10.1 STAGES OF SYSTEMS DESIGN

(Prior to the start of designing the data processing system, the systems analyst has had comparatively little opportunity to use his creative abilities. As we have seen, his work up to this juncture has, for the most part, been that of observing an existing situation but not attempting to alter it in any way. Throughout the investigation the systems analyst plays with ideas at the back of his mind, and there is no exact time at which he turns from investigation to design. These ideas, perhaps fanciful for the moment, prompt him to seek further information from the existing system so that he can assess the viability of proposed new systems arising from the ideas.)

Before turning his efforts to full-time design, it is advisable that the analyst look afresh at the objectives and at his assignment brief. A reappraisal of these in the light of information now to hand enables him to decide their relevancy and realism. Any doubts in these respects should be eliminated by further discussions with top management before using the objectives and the assignment brief as guidelines.

The task of designing each individual data processing system presents its own special problems and unique features. It is not possible to lay down rigid rules of systems design that apply to all cases. Moreover, it is dangerous for a systems analyst to overplay his experience by attempting to impose previous solutions on present problems. Each new situation must be considered against its own background, and the emphasis in the systems design weighted accordingly.

Aspects of Systems Design

When designing a data processing system, the main aspects to be considered are: cost, efficiency and accuracy, practicality, and flexibility.

These factors cannot be rated once and for all; in one company the over-riding aspect is cost minimization; in another it may be a combination of accuracy and practicality. Even within one company, the ratings can change with changing circumstances so that the weight given to the different factors in one year might not hold a few years later. Nor can they be treated as isolated and distinct aspects of design; they are irrevocably connected, a change in any one resulting in some degree of change to each of the others. The weight that the systems analyst gives to each of these aspects must be decided as a result of his knowledge of the company's activities colored by an intuitive forecast of future trends.

Data Processing Routines and Processing Runs

Each application to be computerized almost certainly consists of a number of routines. It is likely that these routines will remain substantially similar in purpose after computerization as when performed by the existing method. A data processing routine is a piece of computer work that achieves results usable outside the system. An application is made up of a number of routines, each of which consists of several processing runs, as shown in Fig. 10.5. The actual numbers of routines and runs forming an application differ considerably, depending upon the type of computer, the system adopted, and the nature of the application. A typical data processing application—sales accounting—might consist of three main routines: invoicing, ledgers, and sales analysis. The invoicing routine might in turn comprise several processing runs shown in Fig. 10.5. The nature of computer-based data processing is such that the routines are interconnected with one another through the use of computer files. The runs are interconnected both through the files and the input/output media, the output of one run very often being the input to another.

In the simplest arrangement a processing run is controlled by one computer program and is performed from start to finish without interruption except for control and security purposes. When implemented on a larger computer, however, it is the operating system and the supervisor program that dictate the precise minute-by-minute scheduling of these activities. As we have already seen, these facilities do not affect the designing of systems to any great extent, and so the designer need not greatly concern himself with such niceties. It is, however, important that the systems analyst give careful thought to the composition of processing runs, and any doubts about the intended capability of a run must be dispelled before building it into the system. If necessary, the systems analyst discusses these factors with an experienced programmer; in an extreme case, it may be advisable to write the program and to test-run it in order to verify its capabilities.

The particular elements of a processing run that must be taken into consideration are:

1. *The approximate amount of core storage needed to hold the computer program's instructions.* With a large computer the amount of core is unlikely to be exceeded, and if the computer is operating in multiprogramming mode, the core is allocated dynamically as and when required for each program so that it cannot effectively be exceeded. A small model, such as a visible record or minicomputer, may have severe restrictions in this respect, however. The salient point at issue is the amount of core storage that is available for holding both data and the program at the one time. If insufficient, then extra time is taken in transferring program segments and data to and from auxiliary storage.

2. *The availability of peripheral units, especially magnetic tape units and line printers.* The designer must clearly not exceed the numbers of these peripherals available on the computer. Generally speaking, a computer run utilizes only the one printer, since the printed output is of only one layout at a time; for example, pay slips, invoices, and statements.

3. *The amount of disk file data needed by the run,* bearing in mind that this is made available either by loading removable disks onto the units or by transferring data from magnetic tape to fixed disks. In any case, the amount of disk storage must be sufficient to accommodate all file data needed at one time during the run.

4. *The sequence and mode of access of the file records* as compared with the sequence of the input data or output results. The particular arrangement of a file may well be decided by the needs of the longest processing run using the file, and so the run in question may have limitations in terms of file access.

5. *A degree of flexibility* must be built into the structure of the run's program. This means that it can cope with any reasonably foreseeable increases or changes in the items dealt with by the run. Of particular importance in this respect are tables of figures, either prestored or created during the run.

6. *The need to detect and deal with errors and exceptions.* The extent to which this is needful depends largely upon the run's position within the overall system, those runs closer to the input of source data being more implicated in error detection. In most cases it is satisfactory for the run merely to indicate the error or exception condition and to continue processing the other data. In dealing with errors and exceptions, it is quite in order for them to be handled manually, provided the procedure is properly linked

to the data processing system and, as far as possible, subsequently checked by it.

7. *The control and audit features* to be incorporated into the run (Section 11.4).

Run Splitting. If any of the above conditions cannot be accommodated within one run, it must be split into two or more runs. In other words a fresh program or program module has to be loaded into core storage to replace the previous run's program. When deciding the point at which to make the split, the following points are worth bearing in mind:

1. When using magnetic tape, it is inefficient to write out large files in the first part of a split run and then to read them again in the same sequence in the second part. This procedure is sometimes unavoidable, but it is generally possible to minimize the size of such files by summarizing the data before writing it onto tape.

2. It is often convenient to split a run involving a final printout at the point just before the printout. The final printout program is usually not connected with the previous processing, and can therefore be loaded just prior to use. This method is of real benefit only if the final printout is of significant complexity.

Source Document Design. When designing a new system, the systems analyst will probably be faced with the task of creating new forms. Two types of forms or documents are involved: source documents and output documents. Source documents are those from which the computer's input is derived; generally these are already in existence for other purposes and consequently must be accepted as they stand for use as punching documents. New source documents should be laid out to facilitate punching, and it may be possible to have the card column numbers actually printed alongside the relevant data on the document. When the new source document is intended solely for computer input purposes, it can be designed with definite positions shown on it for each digit or character to be punched.

Whatever the source document, it is very important that its contents be clear, legible, and definable (Section 10.3). It may be necessary for the systems analyst to insist that the handwriting of entries be improved; otherwise, considerable errors may occur due to misreading by the punch operators.

Output documents are covered in Section 10.5.

10.2 COMPUTER INPUT FROM PUNCHED CARDS AND TAPE

The advances in computer technology during the past 20 years or so have been led by the improvements in electronic circuitry and components. Following behind these have been the developments in storage devices and peripheral units, and in a poor third place have been the methods for creating computer input media. Only in the past few years have these methods started to reach a comparable stage of development. There are now wide ranges of methods and devices available for converting printed or written data into media that are readable by computers. One of the more familiar signs of this development are the stylized figures that appear along the bottom edge of bank checks. Another aspect of computer document reading, and one that has changed our ideas about form filling, is the employment of marks (small horizontal lines) instead of conventional figures on documents. This is known as "optical mark recognition," or mark reading, and is dealt with in Section 10.4. Other devices such as badge readers, tag readers, and on-line terminals are coming into greater usage within their own spheres.

Nevertheless, in spite of these newer developments, the majority of computer input data is still punched manually into cards or paper tape, and it is with these media that we shall deal first.

Punched Cards

Punched cards have been in use for many years in one style or other, but have by now settled down to two main types. These are the long-standing 80-column, 12-row card, and the newer IBM System/3 card. The punching convention of the former is standardized so that they can be read by all computers, regardless of manufacturer.

The most common method of preparing cards for computer input is still the well-established punching and verifying from source documents. This is covered in Section 10.3, so we will deal first with the other aspects of punched cards.

Card Pulling Files. Instead of manually punching a batch of cards to represent movements that occur in the company, cards can be pulled (picked) from a prepunched "pulling" file. This method is well proved and is effective for coping with a large amount of source data that could not otherwise be handled in the time available. The pulling file is replenished by reproducing (using a punched card reproducer), manual punching, or reinsertion of the original cards during off-peak work periods. A pulling file contains several cards to represent each item covered by the file; typical examples of the ranges of these items are:

Commodities in a selling range.
Items held in factory stores.
Constituent parts within manufactured products.
Employees on the payroll.

The number of cards in the pulling file for a given item depends upon its rate of usage; there must be a sufficient number to avoid running out before the next replenishment.

Card per Unit Files. An alternative arrangement to card pulling is for the file to hold cards that represent individual items actually in stock; this is known as a "card per unit" file. Fifty cards in the pulling file for an item means that a quantity of 50 of that item is actually in the stores. This system is disappearing with the increasing employment of computer files, but it does have the advantage that the stock position is immediately apparent, provided an item is withdrawn from stock only against a card. In situations where orders and inquiries are received by telephone, this arrangement can be the most convenient for maintaining a visible, up-to-the-minute representation of the stock position.

Consumable/Nonconsumable Pulling Files. In order to facilitate pulling, the cards are filed in open racks with labels to indicate the item code numbers in the adjacent cards. The cards are generally filed in code number sequence and, as a rule, "interpreted" along the top edge so that it is easier to read their contents. In addition to prepunching of item code numbers and other indicative data, it is sometimes possible to have prepunched quantities in the cards. For each item, there are then a few cards for each quantity that could be involved in a movement or transaction. For example, with an "orders received" file, if the orders are generally for singles of a certain commodity, then most of this commodity's cards would be prepunched with "1" in the quantity field.

The quantity fields often hold small quantities such as 1, 2, 4, 8, or other groups that predominate in usage. A movement involving an uncommon quantity is handled by pulling two or more cards. If larger and more widely varying quantities are in use, the pulled cards are punched manually with the quantity after being pulled. It is also normal to punch other data manually into the cards after pulling; this means that after use the cards cannot be reinserted into the pulling file; this is then a consumable file. When no further data is punched into the pulled cards, they are generally returned to the pulling file after use and reinserted into the correct position therein. This type of file is nonconsumable, but of course the cards must be replaced when they show signs of wear. If a high-speed card reader is in use, it is wise to allow only one reading of the cards; i.e., have a consumable file.

Card pulling has the following advantages:

1. Errors due to the misreading of code numbers and descriptions on documents tend to be reduced.
2. Pulling is faster than punching; therefore, input data can arrive at the computer earlier, or greater volumes can be handled in a given time.

3. The punching load is reduced by a nonconsumable file, and is more evenly spread if the file is consumable.

The disadvantages of pulling files are:

1. The card is inherently tied to one and only one item; this means that "spread" cards (holding several items, and described later in this section) cannot be used. As a consequence, more card-reading time is needed on the computer.
2. Card consumption tends to be higher than with conventional punching methods; even a nonconsumable file cannot compete with spread cards in this respect.
3. The pulling file occupies a considerable amount of space, involving a large room if a wide range of items is represented.

Dual-Purpose Cards. These are punched cards that are also used as documents in that they have handwritten entries made on them; the entries are then hand-punched into the card. A common usage of dual-purpose cards is as plant job tickets. In this case the cards are punched and interpreted with details of the job to be done; after its completion, further details of the job are handwritten on the card, and it is returned for this data to be hand punched and then processed. If dual-purpose cards are to be read by a computer after they have been handled in a factory or office, great care must be taken to insure that they are not damaged. The damage includes creasing, tearing, fraying, and dirtying; if the damage is unavoidable, the cards should be reproduced and new copies read by the computer. A reproducer can accept a far higher level of damage than can a computer's card reader.

Dual-purpose cards are also used in conjunction with pulling files. In this case the item's predetermined data, such as its code number and price, are in the card when pulled. Further data—say, the order quantity—is written on the card after pulling and subsequently punched.

Spread Cards. These cards hold data pertaining to more than one item, and usually involve from 2 to 20 items. Each item's data may be completely self-contained so that, in effect, the spread card is equivalent to a pack of smaller cards joined together. Alternatively, the spread card may hold certain fields whose contents apply to all the items in the card. When a spread card holds a variable number of items, as it very often does, it is equivalent to a short length of paper tape. It is advantageous to be able to fill a card in this way because the card-reading speed of a computer is independent of the card's contents, and it also increases the manual card-punching speed per item. Spread cards cannot be sorted off line (by a mechanical sorter) into the sequence of the item's keys because of the conflict of the various keys therein.

A typical usage of spread cards is the punching of customers' orders, with the ordered items spread across the cards and preceded by the indicative data such as account number, order number, and date.

Paper Tape

There are now two main types of paper tape employed with computers. These are the eight-track (or channel; seven data bits plus parity bit) ISO code, and the seven-track (six data bits plus parity bit) code, both of which are punched in 1 inch wide tape. Most paper tape readers can, in fact, read five-, six-, seven-, or eight-track tape, and any tape coding can be translated by program into the computer's internal representation. With eight-track tape, the seven data bits can represent 127 different characters, digits, or symbols, and—provided the computer's internal representation is the "byte" form (eight bits)—no modification to the code layout is necessary in the computer.

Parity Bits. In seven- and eight-track paper tape, each character is represented by a set of holes across the tape (a frame). A means of checking automatically that no hole has been omitted nor an extra hole inserted by the punching device is to include a "parity bit" in each frame. Parity may be either odd or even, but must be consistent within the one reel of tape. With even parity, the parity bit is inserted in order to make the total number of bits (i.e., holes) in each frame into an even number; with odd parity, the total is an odd number.

Sources of Paper Tape. It is quite usual to punch data into paper tape by means of key-operated punches and verifiers in a manner similar to that used for punched cards (this method is covered in Section 10.3). There is, in addition, its creation as a by-product of other operations; various types of keyboard accounting machines can be fitted with attachments to provide punched paper tape output at the same time as performing their normal functions. This by-product source of data is of particular benefit to small companies wishing to use a computer service bureau for data processing work such as sales analysis and costing.

By-product tape does, however, introduce a number of additional problems not applicable to direct punching. It is most important that during the planning stage the systems analyst maintain a close liaison between the department preparing the tape and the programmers responsible for programming its translation. Thorough consideration should be given to the following factors:

1. *Check digits:* What fields are to contain check digits, and what weightings and modulus are to be employed (Section 5.5).
2. *Special symbols:* The precise meanings and positions of these in con-

nection with field sizes, record sizes (Section 6.2), and error correction and notification.

3. *Error correction procedures:* These may be quite complex because the contents of by-product tape are largely decided by the printed output of the keyboard machine. Any erroneous printing is inevitably duplicated in the tape, so that the operator cannot merely correct the hard copy. Arrangements must also be made to cancel the effect of the erroneous tape when it is read by the computer; this can be done by one of the following methods:

(a) Inserting a symbol after each record to indicate its correctness or incorrectness. If the latter, the correct version of the record is usually made to follow immediately after. To do this, the keyboard operator switches off the punch when correcting the hard copy and then switches off the printer when repunching the paper tape.

(b) "Contra entries" effectively cancel the errors by providing further records in the tape that contain exactly the opposite values to those that are erroneous. A third record is then punched in each case; this contains the correct value.

(c) Inserting correction entries with plus or minus designations, thus eliminating the effect of the errors when added into totals.

Comparison of Input Speeds of Punched Cards and Paper Tape

In systems where large volumes of input data are to be read by a computer, the comparative input speeds of these two media are of considerable interest. The comparison is somewhat confounded by (1) the maximum speeds of the various readers, and (2) the packing efficiency of the media.

With regard to the second point, there are a number of factors worth remembering. First, card-reading speeds are unrelated to the amount of data in each of the cards. Second, paper tape reading time has to allow for the gaps between blocks of data and for any special symbols for field, record, or error identification. Third, paper tape has the advantage of being able to hold variable-length fields. These are not used with punched cards, although there is no theoretical reason against this, so that some of the card columns hold nonsignificant zeroes, and these contribute no real information to the computer. Paper tape can, of course, also contain fixed-length fields in the same way as punched cards. Fourth, when records in punched cards are greater than 80 characters, they must in effect be split into two records, entailing duplication of all or some of the indicative data.

Taking all things into consideration, a high-speed card reader and a high-speed paper tape reader run a neck-and-neck race in the "computer input handicap." This can be seen by reference to Fig. 10.1 and the following examples.

CARD INPUT — CHARACTERS PER SECOND				
Card cols. occupied	Packing efficiency	Card reader speed c.p.m.		
		300	600	900
20	25%	100	200	300
40	50%	200	400	600
60	75%	300	600	900
80	100%	400	800	1200

PAPER TAPE INPUT — CHARACTERS PER SECOND		
Packing efficiency	Paper tape reader speed	
	300 chars/sec	1000 chars/sec
60%	180	600
70%	210	700
80%	240	800
90%	270	900
100%	300	1000

$$\text{Packing efficiency} = \frac{\text{Significant characters per record}}{\text{Columns or frames per record}}$$

$$= \frac{\text{Card columns occupied}}{80} = \frac{\text{Significant characters}}{\text{Sig. chars} + \text{symbols} + \text{gaps}}$$

Fig. 10.1 Punched card and paper tape input speeds

Example A. It is required to read into the computer 30,000 records of 60 characters. Punched into cards this represents a 75 percent packing efficiency, and when using a 900 card per minute card reader, the time taken is

$$\frac{30{,}000 \text{ records} \times 60 \text{ characters/record}}{900 \text{ characters/second (Fig. 10.1)}} = 2000 \text{ seconds} = 33.3 \text{ minutes}$$

If punched into paper tape with five symbols per record and a 1-inch gap between records, the packing efficiency becomes

$$\frac{60}{60 + 5 + 10} = 80 \text{ percent}$$

When employing a 1000 character/second paper tape reader,

$$\text{Time taken} = \frac{30{,}000 \text{ records} \times 60 \text{ characters/record}}{800 \text{ characters/second (Fig. 10.1)}}$$

$$= 2250 \text{ seconds} = 37.5 \text{ minutes}$$

Example B. Same conditions as in Example A except that there are 30 significant characters per record; i.e., 67 percent packing efficiency in paper tape. Reading on a 900 card per minute card reader takes

$$\frac{30{,}000 \times 30}{450 \times 60} = 33.3 \text{ minutes} \text{ (50 percent packing efficiency)}$$

and on a 1000 character per second tape reader,

$$\frac{30{,}000 \times 30}{670 \times 60} = 22.4 \text{ minutes}$$

Comparison of Compactness

Paper tape is considerably more compact than punched cards. Taking an example of 30,000 records, each of 60 significant digits, the amounts of space occupied are:

1. *80-Column Punched Cards.* One card per record, each card $7\frac{3}{8} \times 3\frac{1}{4} \times 0.0067$ inches, gives an absolute volume of approximately 4800 cubic inches, or an effective volume of 4 cubic feet when in trays.

2. *IBM 96-col (System 3) Punched Cards.* One card per record, each card being approximately one-third of the volume of the 80-column card, gives a volume of 1600 cubic inches, effectively 1½ cubic feet in trays.

3. *Paper Tape.* A given 7½ inches of tape per record (including five symbols and a 1-inch gap) means 19,000 feet of tape; i.e., 20 reels of 10½-inch diameter. When in containers these occupy an effective volume of 2 cubic

feet. If the cards hold only a small amount of data, the card/paper tape ratio can increase to 8:1.

Media Cost Comparison

Paper tape also has a cost advantage over cards, even allowing for the fact that tape verification entails using twice as much paper tape as is actually needed to hold the data.

The cost ratio of the media for a given annual data volume lies between 7:1 and 1.5:1, and is likely to remain within these limits regardless of rising costs.

Comparison of Flexibility of Use

The well-worn argument that paper tape cannot get out of sequence, whereas punched cards can, is meaningful only in certain circumstances. If there is no other method of sorting except off line, then punched cards must be used anyway. If, on the other hand, the computer can sort on line, as is usually the case, then the sequence of the input is irrelevant. With either medium, it is unwise to rely on computer input as being in exact sequence.

Paper tape can be used in conjunction with data transmission equipment more readily than can punched cards; it is both faster and cheaper for this purpose. It is also more applicable to minicomputers than are punched cards.

Punched cards are more visually readable, especially if interpreted. This advantage allows their use as documents. Cards also have the advantage of being easier to correct; it is obviously simpler to extract one card and replace it by a corrected version than to find and amend a tape record.

10.3 MANUAL KEY OPERATING

This method still accounts for the highest proportion of computer input data, although this situation may not apply a few years hence. Many of the principles involved also apply to writing on data recorders and key stations.

Machine Facilities

Normally, punched cards are prepared by one operator punching a pack of cards, which are then verified by another operator. A variety of punch/verifiers are available on the market, most of which are also capable of interpreting the data in the cards and some of which have a check digit generator fitted.

Paper tape is verified either by reading manually through the hard copy that is produced simultaneously with the punched tape or by comparing the verifier operator's keying with the contents of the primary tape. A secondary tape is produced during this process, each frame being actually punched only if the keying and primary punching agree. The secondary tape is the one used as computer input.

Punching Instructions—Cards

The items on a source document must be legible and identifiable; there is not much that can be done about illegibility except to create the data in some other way or take more care in writing. Identification can, however, be facilitated by the employment of clear punching instructions allied to a specimen source document. For each type of source document a "punching instructions" form should be filled in at the time when the related card layout is first designed. Thereafter, this form is filed in the punch room and referred

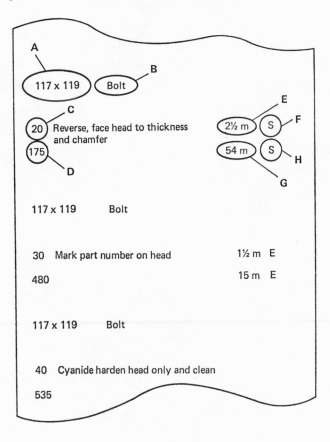

Fig. 10.2 Example of a source document

PUNCHING INSTRUCTIONS FORM

Department				Queries to		
Production control				Mr. Underwood		

Name of source document				Ref. No. of source document		
Operations master				—		

Name of card			Electro No.	Color stripe	
Operations master			4-625	Blue	

Ref.	Field name	Card cols.	Picture	Punch left or right	Fill with	Remarks
	Class of card	1	9	—	—	= 6
	Serial no.	2	9	—	—	Punch sequentially within each part no.
A	Part No.	3-9	999A999	Right	Blanks	
B	Description	10-19	A (10)	Left	Blanks	
C	Operation No.	20-21	99	Right	Zeros	
D	Machine No.	22-24	999	Right	Zeros	
E	Operation time	25-27	99·9	Right	Zeros	Punch halves as 0.5
F	Operation desgn.	28	9	—	—	Punch E as 1, S as 2
G	Set-up time	29-31	9·99	Right	Zeros	Hours and mins.
H	Set-up desgn.	32	9	—	—	As F
C-H	2nd operation	33-45				As cols. 20-32
C-H	3rd operation	46-58				As cols. 20-32
C-H	4th operation	59-71				As cols. 20-32
		72-80	B (9)	—	—	

Notes:

Ref.   Letter refers to field of attached specimen of source document

Picture   A = alphabetic field, 9 = numeric field,
X = alpha-numeric field, B = blank
9 = numeral 0 to 9

Fig. 10.3 Punching instructions form

184

to whenever there is any doubt about the layout of a card. An example of a filled-in punching instructions form is shown in Fig. 10.3; its entries relating to the specimen source document are shown in Fig. 10.2. The form in Fig. 10.3 covers most requirements but can, of course, be modified to suit a particular user's needs. Its contents are:

1. "Department": where the documents originate.
2. "Queries to": the name(s) of the person(s) who can help with queries regarding the content of a difficult document.
3. "Name" and "Reference Number" (if any): identifies the source document.
4. "Name" and "Electro Number" (or equivalent): identifies the card; the card name is often the same as the document name.
5. "Color Stripe" on card (if used).
6. "Reference": see note on form.
7. "Field Name": should correspond to that printed on the document and/or card.
8. "Card Columns": first and last columns occupied by the field at its maximum.
9. "Picture of Field": see notes on form.
10. "Punch to Left or Right": this indicates the method for accommodating unfilled fields. Numeric fields are normally "justified" right, i.e., with nonsignificant zeroes at the left of the field; alphabetic fields are normally "justified" left, with unfilled columns remaining blank to the right.
11. "Fill with Blanks or Zeroes": refers to the unfilled columns of the field.
12. "Remarks": usually unnecessary, but any additional instructions can be inserted here.

Class of Card (card type or designation). This is a one- or two-column field punched into every card so that both the computer and the staff can identify its purpose. Although, as far as the computer is concerned, only cards used together in the same processing run need be differentiated, it is good practice to have a unique "class of card" for each type of card used in the data processing department. This means that several score, if not a hundred or more, classes of card are needed to meet all the various input formats of business applications. The coding in the "class of card" field may be both alphabetic and numeric, giving 38 different codes if one column is used, or 1444 if two columns. The same columns should be used for "class of card" throughout all cards; generally, columns 1, 1 and 2, 79 and 80, or 80.

Punching Instructions—Paper Tape

The instructions for paper tape punching are similar to those for cards, but exclude name and electro number, color stripe, and card columns. Additional instructions are necessary with regard to field symbols for variable-length fields, record symbols, error correction entries, and tape heading labels used to identify the piece of tape.

It is advisable to leave gaps in the punching at intervals along the tape, say, every 6 inches or between records, whichever is longer. The gaps facilitate finding particular records and also indicate insertion and correction of records by splicing.

Timing of Punching Operations

When estimating the time needed for the preparation of punched cards or paper tape, an average figure of 10,000 key taps per hour may be used. This figure is general, and in practice is modified by the following factors:

1. The legibility of the source documents, especially if these are hand-written.
2. The layout of the source document.
3. The number of queries raised; usually these are results of illegibility or missing entries.
4. The layout of the punched medium, including the gaps and symbols in paper tape, the spaces in cards, and the proportion of the card occupied by data.
5. Verification requires another 10,000 key taps per hour.

For each type of card and paper tape, it will be found from experience that a certain speed of punching can be maintained. In the planning stage the general rate of 10,000 key taps per hour, plus an allowance of 1 second per card for feeding time, is satisfactory. If it is known that the source documents will be of poor quality, an additional time allowance should, of course, be made.

Thus, if the daily punching load is 3000 cards in which an average of 60 columns are punched, the estimated time is

$$\frac{3000 \times 60}{10,000} \text{ hr} + 3000 \text{ sec} = 18 \text{ hr } 50 \text{ min}$$

This means that almost 19 machine hours for punching and the same for verifying must be allowed. If this work load has to be completed during 5 hours of the working day, then at least four punches and four verifiers together with eight operators will be needed.

The preceding example is obviously an oversimplification of a real situation. The punching load is likely to be made up of many different batches of cards fluctuating in size from day to day. Nevertheless each significant batch should have its punching time estimated so that these can be fitted into the scheduling of routines (Section 11.3).

Magnetic Tape Encoders and Key Stations

It has been realized for many years that the manual punching of cards and paper tape is a slow and cumbersome method of preparing data for input into a computer. Card and paper tape punching and reading speeds are inordinately slow as compared with other means of inputting data, such as magnetic tape. During the past few years there has been a gradual increase in the employment of this latter medium as the prime means of entering source data into a computer.

Depending on the particular manufacturer, the equipment that is employed for preparing the magnetic tape from source data is known by various names such as magnetic tape encoder, data recorder, and magnetic data inscriber. Essentially, all have a keyboard, an illuminated display, a buffer store, and a magnetic tape drive. The buffer store is necessary to balance the high speed at which the device writes data onto the magnetic tape against the relatively low speed of the manual keying. The same unit is used for both recording and verifying, and it also has the ability to search for any particular data record that the operator wishes to inspect.

In operation, the data is keyed into the buffer until it is full. This block of data is then written into magnetic tape, which is then automatically backspaced for checking purposes. When verifying, the data on the magnetic tape is read into the buffer store and then compared with the same data keyed in a second time; any disagreement causes the machine to lock until the error is reconciled. The data recorded by individual operators on short lengths of magnetic tape is automatically pooled onto one reel or, alternatively, fed into one multiplexer channel before reaching the computer.

The recording of data directly onto magnetic tape has several advantages: First, the elimination of the slow reading of punched cards and paper tape; second, the increased speed of keying due to the improved quietness and smoothness of this operation; third, the avoidance of continual punched card costs, since magnetic tape can be used repeatedly.

A further development along these lines is the "key station" system. This consists of a number of keyboards (about ten or more), all connected to one small, special central processor. The data from the keyboards is automatically pooled and written onto a disk or drum storage device. After being sorted and edited, the data records are then transferred onto magnetic tape or disk for input to the main computer.

The aforementioned devices are, of course, considerably more expensive than conventional card or paper tape punches, but are nevertheless worthy of consideration if large volumes of source data are handled.

10.4 NONKEYED INPUT MEDIA

Magnetic Ink Character Recognition (MICR)

The MICR system has been developed as a means of providing printed figures that can be read by both humans and machines. By printing the figures in magnetic ink and in a stylized form (font), this dual role becomes feasible. There are two MICR fonts: the British and American standard is the E13B font, as is used by the banks. The E13B figures can be seen across the bottom of bank checks, and in addition to the stylized numerals 0 to 9, four special symbols are employed to signify the meanings of the fields.

The Continental standard MICR font is known as CMC7, and differs from E13B in appearance. The range of CMC7 characters consists of the numerals 0 to 9, the alphabet, and five special symbols. Each character is formed from seven vertical bars whose length and vertical position are arranged to form the character's appearance, and whose horizontal spacing enables the MICR reader to recognize the character. The size of the characters is approximately the same as the E13B characters.

The use of MICR documents for purposes other than as bank checks is limited by the size restrictions imposed by MICR readers. The documents must be approximately 6 inches wide by 3 inches high, and the stylized characters must be 0.25 inch from the bottom edge.

The original purpose of magnetic ink characters was to facilitate the high-speed sorting of documents when returned for accounting procedures, but it is now possible to use MICR on line and thereby feed their data directly into the computer. The MICR sorter/readers operate at speeds of up to 1600 documents per minute, each document being encoded with a maximum of 75 numerals. This gives a theoretical maximum input speed of 2000 characters per second; this is higher than the speed of paper tape and card reader inputs.

The magnetic ink is unaffected by being overwritten by ordinary ink, and the sorter/readers have a high degree of tolerance toward dirty and damaged documents; moreover, the nature of the ink makes forgery of documents difficult. However, owing to the restricted means of encoding MICR documents and their size and reading limitations, they have not become widely used in business generally. The advances in optical character recognition have made this a more suitable system for the majority of commercial and industrial applications.

Optical Character Recognition (OMR and OCR)

This subject involves three different types of optical readers: mark readers, document readers, and page readers. The first two of these devices are sometimes combined to form one unit, called a "universal document transport."

Mark Reading (OMR). A mark reader can optically detect small black lines made on a document. The lines are roughly 0.25 inch wide, spaced 0.20 inch vertically, and have from 16 to 24 such marks on one line.

The marks can be made in a number of ways:

1. Manually, using a soft lead pencil or a repro-graphic pen.
2. By embossed plates; ordinary printing may also be done simultaneously from the same plate.
3. By a computer printer, printing hyphens.
4. Preprinted when the documents are originally printed.

When marking the documents manually, conscious mistakes are eradicated by extending the erroneous mark downward so as to form a small rectangle; the mark reader then ignores this type of mark.

Documents are read at speeds of up to 1200 documents per minute, but the actual speed attained depends upon the size of the documents being read. The precise document sizes that can be handled depend upon the particular model of universal document transport, but all models accept a wide range of sizes up to about 12 X 8 inches. An important feature of mark reading is the flexibility of document layout; this results in versatile applications of the system. The marks do not have fixed values, meanings, or positions on documents; their interpretation is entirely by computer program.

Among the very wide range of existing applications of mark reading are the following:

1. *Customer order forms:* Forms are printed with rows or columns of commodity names, with spaces for marking the order quantity alongside. The customer's account number, week number, etc., may be printed by the computer as marks at the top of the form before sending it to the customer. This arrangement provides an almost foolproof ordering system because the minimum of manual marking is required. An elaboration of this principle is to produce the order document in the form of a diagram that shows, for instance, an exploded view of an assembly or product. Each component is arbitrarily numbered on the diagram and the required component is indicated by putting a line through the appropriate number.

2. *Meter reading:* These forms are generally preprinted with the consumer number in the form of marks or optical characters.

3. *Survey questionnaire sheets:* These must be very simple and easy to mark if they are intended for use by the general public.

4. *Time sheets:* These are used by many large organizations.

5. *Examination answer sheets:* These are suitable only for questions that have a limited number of prestated answers from which a choice must be made.

Document and Page Reading (OCR)

A document reader is limited to reading only one or two lines of OCR characters per document. A page reader can read a full page of such characters. The main fonts employed are:

OCR "A": American Standards Association.
OCR "B": European Computer Manufacturers' Association.

The OCR "A" has 56 different letters, numerals, and symbols; OCR "B" has no less than 113. Both fonts have four standard character sizes. These OCR documents lend themselves readily to certain applications and particularly to those that can make use of turnaround documents. These are printed in OCR font by the computer, are sent out for external use, and then returned for further processing by the computer. This system has tremendous advantages in that the indicative data (such as names and addresses, account codes, or part number) is absolutely accurate because it does not have to rely upon people's memories or manual transcription from other documents.

The comparative slowness in the general adoption of this method is undoubtedly due to its expense. If this cost problem can be overcome, either by cheaper hardware or the employment of service bureaus, there is every reason to believe that it will be very widely adopted. Combined with mark reading as a means of reprocessing variable input data, OCR documentation could well become a standard procedure in industry and commerce.

Tag Reading

A tag is a small card attached to goods or parts mainly for the purposes of stock and production control. It is sometimes divisible into sections that can be detached during the handling of the item concerned. The encoding of the tags includes small holes, typewritten 1's, and handmade marks. A tag reader operates at up to 500 tags per minute, each tag holding about 15 digits to accommodate code number, price of item, and similar data.

Badge Reading

Badges are embossed plastic cards read by off-premise terminal stations; the data is converted into paper tape at the central receiver together with manually operated switched data. This on-line terminal input has a variety of applications.

On-Line Data Input. There are obvious advantages to be gained from the direct input of data to the computer. That is to say, without using any intermediate medium or equipment other than a keyboard or visual display unit. On-line input involves the employment of terminals from which data is transmitted into the computer, and at which results or comments are subsequently received. Increasing use is being made of terminals as the means of prime data collection, and this has the advantage of early detection of errors of certain types. In particular, nonexistent items are immediately detected and notified because the computer cannot find these on its files; thus, the terminal operator can be prevented from continuing to key-in that particular transaction.

Where the terminal comprises a visual display unit, this can be used to guide the operator by displaying alternatives from which the input data must be chosen. This would apply, for example, to an ordered item having a limited number of optional characteristics such as size or color. In a similar way, a user can be guided in obtaining information from files through a file-interrogation procedure.

Real-Time Data Input. The employment of terminal input-output is an absolute necessity with real-time business systems, since by their very nature these must be up to date. The leaders in the real-time field are undoubtedly the airlines because these companies must be able to accept airline seat inquiries and reservations on an interactive (conversational mode) basis from visual display terminals situated in all parts of the world. Owing to the immediate response of such systems, and since the data is collected very close to its source (i.e., the prospective airline customer), error conditions and misunderstandings are largely obviated. Similarly, the ability of all terminals to access and update the central flight reservation file avoids the possibility of overbooking and thus reduces the tendency to accept duplicated bookings.

Bank and savings and loan systems are less conversational because the transactions are more stereotyped, but nevertheless the input data (monetary transactions) immediately updates the central accounts file; consequently, overdrawals and fraud can be reduced.

In addition to keyboard input, real-time and on-line systems can be arranged to accept input data of a more analog nature, such as graphs or drawings by means of a lightpen. Although this procedure is now well used for technical and scientific purposes, it is not likely to extend deeply into business applications, owing to its expense and only the occasional need for it.

10.5 DESIGN OF PRINTED COMPUTER OUTPUT

The documents printed by the computer and subsequently distributed to the user departments or to outside agencies are very often the only indication that the data processing department exists, as far as they are concerned. The printed document is tangible evidence—but evidence of what? A badly designed layout or confusing terminology associated with it are interpreted as inadequacy. As a result it is heavily criticized even though the information thereon may be fully correct.

The systems analyst, in cooperation with stationery manufacturers and the user department concerned, should go to great lengths to insure that maximum clarity is achieved on output documents. This is especially necessary when the document is intended for outsiders such as customers, and also when it is replacing another document that has maintained a high standard of printing such as that produced by an electric typewriter. We must remember that the likely absence of small letters (lower case) and the not-too-high standard of computer printing (especially if multiple copies are produced) may cause the document's information to be less easy to read and difficult to understand.

The proper way to design computer output is to work backward from what is the most desirable layout from the users' point of view. It is not satisfactory to approach the problem from the other way; i.e., to design the printed layout merely to suit the programming, the computer printer's characteristics, or the file layouts, although these factors do enter into the problem to some extent.

The designing of the layout of printed output is facilitated by the use of print charts supplied by the computer manufacturer. These are simply blank forms divided by vertical and horizontal lines to indicate the spacing of the printing and to allow for entries being made by hand. Since the print charts are the full size of computer printer spacing, the manual entries are a replica of the desired computer printout. The usual spacings are ten characters per inch horizontally (including blanks), and six lines per inch (or a submultiple of this) down the sheet.

The factors to be taken into consideration in designing the layout of computer printing and the preprinted documents are as follows:

1. *The information to be shown on the document* and its relative positions thereon in order to facilitate easy recognition and understanding of the contents. Vertical and horizontal spacing are especially pertinent to legibility.

2. *The positioning and content* of either preprinted or computer printed headings. The former can be styled to suit the space available; the latter may have to be abbreviated, since headings tend to be wider than the columns of figures beneath them.

3. *The need for titles,* serial numbers, and/or page numbers on each sheet.

4. *The physical size of the document* in relation to the width of the print bank; this is normally from 10 to 16 inches, corresponding to 100 to 160

characters. It should also be borne in mind that the document size may be determined by external factors such as the need for enveloping, stationery suppliers' limits, use in bound form for reference purposes, and so on.

5. *The desirability of allowing blank areas* on output sheets. These areas may be necessary in order to avoid splitting a tabulation, list, or analysis between two sheets when it is better accommodated on one sheet.

6. *The number of copies required* and any variations in contents of the multipart copies; i.e., the need for print-through, blacked-out, or blanked-out areas.

7. *The legibility of computer printing,* especially on colored sheets and if multiple copies are produced. It is generally considered that the best colors are white, pink, pale green, and primrose, in that order.

8. *The employment of computer output* as turnaround documents for subsequent reading by document readers. This can involve preprinted and/or computer printed OMR marks and/or OCR characters.

9. *The possibility of adopting spread printing* (Section 11.2) as a means of increasing printed output rate.

10. *The cost of stationery* of a complicated nature. It is well worthwhile discussing any unusual requirements with the stationery printers or suppliers before embarking upon a definite plan. Costs may be higher than can be justified, and simplifications may result in considerable savings in cost.

10.6 FLOWCHARTING OF SYSTEMS

This subject has already been mentioned (Chapter 4) in connection with the diagrammatic representation of information flow in the existing system. When only manual operations are included in a system, it is convenient to employ a simple arrangement of boxes to represent the operations (Fig. 4.3).

In systems design, the analyst becomes involved with the representation of data processing systems at various levels. At the highest level, a system flowchart is used to illustrate the interconnection of routines that form the wholly or partially integrated system. An example of this is shown in Fig. 10.4. In this chart there is a box for each routine, together with interconnecting lines. The solid lines are intended to indicate the movement of information within the system; the dotted lines indicate the connection between routines via the system's environment. Each routine should be clearly named, and it is helpful to code each one for reference purposes.

The next lower level of flowcharting involves a flowchart or set of flowcharts for each routine. This shows the interconnection of computer processing runs and other operations within a routine (Fig. 10.5) and utilizes a set of special symbols. The shortened set in Fig. 10.6 is extensive enough for

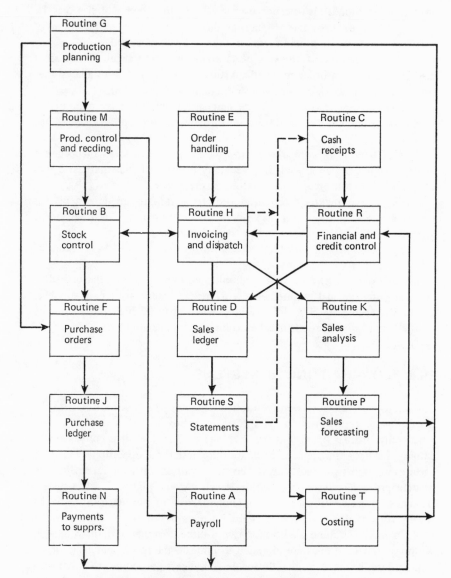

Solid lines represent movement of data between routines, broken
lines represent data movement via outside agencies

Fig. 10.4 System flowchart

most purposes; these symbols (simplified in a few cases) conform to the
recommendations of the International Organization for Standardization. They
are also accepted by the computer manufacturers. The full set of symbols is

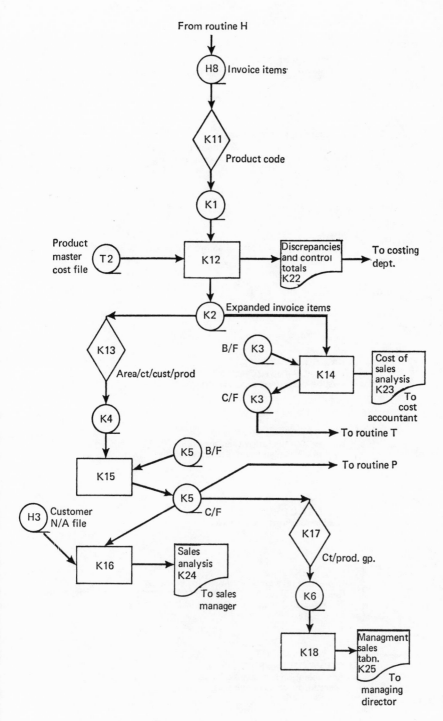

Fig. 10.5 Routine flowchart

195

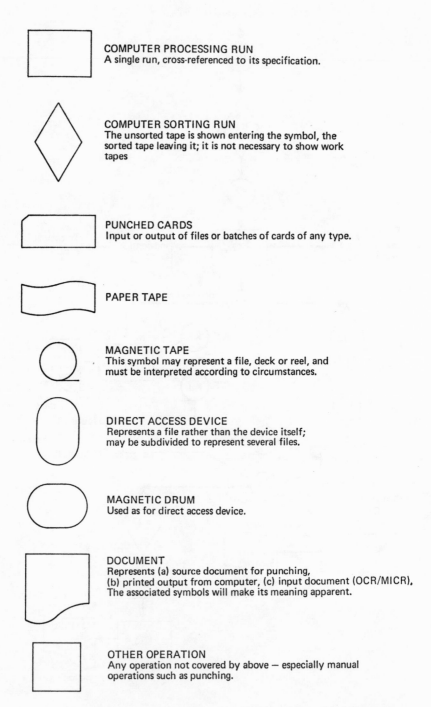

COMPUTER PROCESSING RUN
A single run, cross-referenced to its specification.

COMPUTER SORTING RUN
The unsorted tape is shown entering the symbol, the
sorted tape leaving it; it is not necessary to show work
tapes

PUNCHED CARDS
Input or output of files or batches of cards of any type.

PAPER TAPE

MAGNETIC TAPE
This symbol may represent a file, deck or reel, and
must be interpreted according to circumstances.

DIRECT ACCESS DEVICE
Represents a file rather than the device itself;
may be subdivided to represent several files.

MAGNETIC DRUM
Used as for direct access device.

DOCUMENT
Represents (a) source document for punching,
(b) printed output from computer, (c) input document (OCR/MICR).
The associated symbols will make its meaning apparent.

OTHER OPERATION
Any operation not covered by above — especially manual
operations such as punching.

Fig. 10.6 Flowchart symbols (simplified set from USASI range)

rather more extensive than those shown in Fig. 10.6, but it is not likely that those missing will be required very much; if needed, the details are readily available from the computer manufacturers. If any variation from these standardized symbols is found to be desirable within a data processing department, the variations themselves should be consistent between all flowcharts. A template, obtainable from the computer manufacturers, is useful when drawing the flowcharts.

The symbols are really intended to illustrate the concepts of the system rather than actual hardware devices. In simpler systems the distinction between these two is somewhat blurred, but this is of no consequence, provided the intention is quite clear. The lines joining the symbols represent the movement of data between processing runs; wherever possible, these should be in one general direction, usually from top to bottom, alternatively from left to right of the flowchart.

Contained within or near each symbol is a code number that cross-references each processing run to its program specification (Section 10.7). As can be seen from Fig. 10.5, it is convenient to prefix the symbol code number with the routine code letter, and also to allocate the numbers in groups. This allows for insertion of additional numbers and helps avoid confusion between the files, runs, and documents. Where a file is common to two or more routines, it always has the same code number, which is generally allocated according to the first routine to use it.

10.7 PROCESSING RUN SPECIFICATION

During the design of the data processing system, flowcharts are prepared along the lines described in Section 10.6. These in themselves are of very little use to programmers, and it is necessary to explain in detail each processing run by preparing a specification; an example is given in Appendix II and several others are presented in the literature.*

The depth of detail included in the specification is dependent upon the programmer's experience and knowledge of the system. In most situations his knowledge is limited to programming only, and the systems analyst should therefore prepare the specification in full detail. This is a good policy in any case because memories are short and good documentation facilitates future amendments to the run. It is extremely aggravating to have to reconstruct a system specification by translating program statements in order to discover what the programs purport to do.

Quite apart from receiving a copy of the run specification, the programmer should discuss it with the analyst and also ask questions about the run as he

*Clifton, *Data Processing Systems Design,* Auerbach, Philadelphia, 1971.

proceeds with the program. The worst situation is when a poor specification is presented to an unquestioning programmer; he is then apt to write the program in the way he vaguely believes to be the intention, only to find at the system-testing stage that the program does not fulfill its real requirements. This sort of difficulty is avoided not only by providing a good specification, but also by the systems analyst's keeping in contact with the programmer during the program-writing phase.

Run Specifications

The four main features of a computer run are its input, processing, files, and output. These break down further into the following groups of information, all of which must be included in a run specification.

1. *General description:* This explains the purpose of the run in relation to the systems flowcharts, and cross-references it to the routine of which it forms a part. Any unusual terminology and special conditions should also be explained.

2. *Input layouts, punched cards:* For each class of card used as input, supply the

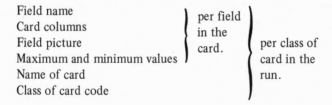

Field name
Card columns
Field picture   per field
Maximum and minimum values   in the   per class of
Name of card   card.   card in the
Class of card code   run.

3. *Input layouts, paper tape:* For each paper tape file used as input, provide the

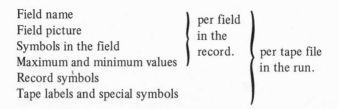

Field name
Field picture   per field
Symbols in the field   in the
Maximum and minimum values   record.   per tape file
Record symbols   in the run.
Tape labels and special symbols

4. *Input layouts, OMR documents:* For each layout of marked documents and for each marked row thereon, designate the

Field name.
Field picture.

Maximum and minimum values of field.

Value of each marking position (cell).

Relationship of cells to one another (for example, are the marks additive?)

5. *Input layouts, MICR and OCR documents:* The position of each line of stylized characters together with a description of each field (as per OMR documents). The meaning of all interfield symbols and special characters.

6. *Volumes of input:* Stipulate the

Average and peak quantities of each class of card.

Average and maximum fields per spread card.

Average and peak numbers of each paper tape record.

Average and peak numbers of documents.

7. *Sequence of input:* Is the input in sequence? If so, what is the procedure if it is found to be out of sequence? Are there any serial numbers to be checked for continuity or completeness?

8. *Meanings of coded fields:* When it is necessary for the program to translate code numbers into a value or a descriptive form, these must be specified; for example,

Factory code,	1 = Buffalo, 2 = Pittsburgh, 3 = Atlanta.
Discount code,	1 = 20%, 2 = 15%, 3 = 10%, 4 = 5%.
Unit of measure,	A = feet, B = square feet, C = pounds.

9. *Stored data volumes:* This applies to items about which data is stored in core, such as tables and indexes. The volumes of these items are helpful in allocating core accommodation.

10. *Processing:* A precise step-by-step explanation of the processes between the input of data and the output of results is required. These processes include all calculations, comparisons, tests, file amendments, etc. Where a calculation is at all complicated, the specification should include a worked example. With multiplications and divisions, the degree of accuracy and the method of rounding off are needed. Decision tables (Section 11.1) should be used to explain complex alternative conditions and actions.

11. *Programming flowchart:* As a general rule, the systems analyst is well advised not to attempt to create a programming flowchart. It is unnecessary and superfluous, since the programmer should be able to do this for himself, and in all probability can do it better than the analyst. If a section of a run is difficult to explain in ordinary language, there is no harm in the analyst's drawing a flowchart as a means of conveying his meaning. This is then carefully explained verbally without his attempting to force the programmer to adopt the precise method; provided the result is correct, that is as much as the analyst should expect.

12. *Error checks:* Details of all feasibility checks (Section 5.4) and check digits (Section 5.5), and the procedures to be adopted if these fail; these may depend not only on the check itself but also on the position of the run when failure is detected.

13. *Control totals:* All control totals that are to be accumulated, with an explanation of any unusual totals. Control totals include both input totals for checking and output totals for printing (Section 11.4). Action is to be taken on finding a discrepancy in an input control total.

14. *File layouts and organization:* For each existing file that is to be accessed during the run, and for each new file created by the run, the details needed are

Number of records, now and in future.
Contents and layout of each type of record.
Mode of storage and sequence.
Key(s) within records.
Magnetic tape block sizes.
Method of addressing direct access buckets.
Organization of cylinders, buckets, and indexes.
Allowance for and organization of overflow.

When the run involves the creation of a new file, it is important that its characteristics be agreed upon by both analyst and programmer, and also that they follow accepted file standards within the data processing department.

15. *Security arrangements:* At what points of the run and under what circumstances dumping is to be performed. What files are to be duplicated (if any), and a description of any other special arrangements for the protection of files or other data.

16. *Output layouts:* Output in the form of punched cards and paper tape is specifiable in the same way as for input.

Printed output (Section 10.5) is most satisfactorily specified by the use of print charts; these are to scale and so facilitate the design of preprinted stationery. When the output is to be on blank stationery, the print layout ought to include the column headings to be printed at the top of each sheet. Other features of printed output are

The number of copies to be printed at a time.
Sheet-numbering method.
Vertical spacing between lines of print.
Detailed rules for continuing to next sheet so as not to divorce totals from their itemized sections.

17. *Sizes of totals:* An indication of the maxima of totals, both intermediate and final, obviates the loss of the most significant digits of a total. This can occur as a result of either overflow from the core locations assigned to a total or failure to allocate sufficient printing positions for it.

10.8 EXERCISES

Problem 1. Input Preparation

The ARCOS Company handles 800 customers' orders per day, averaging 60 items per order. A decision has to be reached whether to use a consumable pulling file or spread cards. The preliminary information required for this decision is (1) the amount of punching and verifying involved, and (2) the card consumption.

If a pulling file is employed, there will be one card pulled per item per order into which a four-column quantity will be subsequently punched. Also, there will be one header card per order into which will be punched the customer's account number (five columns), customer's order number (six columns), and the date (six columns).

If spread cards are employed, the header details as given above will be punched into every spread card together with as many item details as possible. The details appertaining to each item are commodity code (four columns) and quantity (four columns).

Solution to Problem 1. If a consumable pulling file is used, it will be necessary to pull 800 X 60 = 48,000 item cards per day and merge these with 800 header cards. The card consumption is therefore 48,800 cards per day. The number of columns per day of punching and verifying involved is

Items	48,000 X 4 =	192,000
Headers	800 X 17 =	13,600
	Total =	205,600

If spread cards are used, the data punched into each card will be the same header details (17 columns) plus seven item details; i.e., 73 columns in all. Thus, there will be 60 ÷ 7 = 9 cards per order, resulting in a total card consumption of 5600 cards per day. The punching and verifying load will be 800 orders X 17 columns plus 48,000 items X 8 columns, or 397,600 columns per day.

Problem 2. Input Preparation Costs

In relation to the facts given in Problem 1, work out the annual costs of (a) cards, (b) punch/verifier operators, and (c) pulling clerks. Assume that the cards cost $2.00 per thousand, punch/verifier operators are paid $150 per week and average 50,000 columns punched/verified per day, pulling clerks are paid $18 per day and average 6000 cards pulled per day. There are 48 working weeks in the year and 5 working days per week.

Solution to Problem 2. The number of pulling clerks needed is 48,000 ÷ 6000 = 8, and with this method the number of punch/verifier operators is 205,600 ÷ 50,000 = 4 of each (i.e., 8 operators). The spread card method has approximately double the punching load and will therefore need 16 operators but, of course, no pulling clerks. The number of cards pulled and consumed per annum is 48,800 × 5 × 48 = 11.8 million. The number of spread cards used per annum is 5600 × 5 × 48 = 1.4 million. A summary of costs follows:

	Consumable Pulling File, $	Spread Cards, $
Cards	23,600	2,800
P/V operators	57,600	115,200
Pulling clerks	69,120	—
Total	150,320	118,000

It must be emphasized at this point that the costs quoted are not necessarily meaningful in isolation and must be actually associated with all other costs of the two methods. The spread card method is likely to involve more complex hardware because of the need to store more file data (such as prices and sales tax) than with the pulling file, the latter being capable of holding these details in the cards themselves.

Problem 3. Punching Instructions

Fill out a punching instructions form to show a card punched with the data shown in Appendix I. Remember that the data, as it stands, cannot quite be accommodated in an 80-column card, and so it is necessary to reduce the space occupied by "produce." This can be done on the assumption that the computer will subsequently translate the codings or abbreviations of "produce" into their full names.

Solution to Problem 3. This is a straightforward procedure involving first the determination of the longest fields of each type. For instance, "District of Columbia" is the longest state name, and so sufficient columns must be allocated to this field to accommodate this name. As regards "produce," there are a number of methods of dealing with the representation of these entries, such as:

1. It will be observed that there is a limited number of different produce names (18) and so a code number, say, 01 to 18, could be assigned to each type of produce. Since there are never more than six types of produce per state, only 12 columns would be required per card, the codes being punched in any of the six fields. Similarly, if the code were alphabetic (A through R), only six columns would be required.

2. A distinct field could be assigned to each type of produce; i.e., occupy a predetermined fixed position in the card. The minimum size of such fields is one column, the presence or absence of a type of produce being indicated by a "one" or a "blank" in the appropriate column. Thus, 18 columns would be needed for this purpose. The filled-out punching instruction forms are shown in Fig. 10.7.

Problem 4. Flowcharting

Draw a routine flowchart for the stock analysis routine (routine L) described below, inserting code numbers of your own creation into the symbols.

Stock receipts and issues enter the routine in the form of magnetic tapes and are sorted separately into part number sequence. They are then used to update the stock file, which is held permanently in part number sequence, at the same time a printout is created of unmatched items and control totals. Following this, the stock issues are resorted by job number and processed to provide the stock usage analysis.

The updated stock file is read into the computer together with a standard costs file (also held in part number sequence) and a costed stock file is output. This file is sorted into stock classifications before being processed to produce the stock cost analysis.

Solution to Problem 4. See Fig. 10.8.

Problem 5. Processing Run Specification

Draw up a processing run specification for preparing an analysis of the data given in Appendix I. The input is to be in punched cards, laid out according to the second arrangement in Fig. 10.7. The information required from the analysis is

Total area (in square miles).
Total population (in thousands).
Average area of states (to nearest square mile).
A count of the number of states yielding each produce (i.e., 18 counts).
Full details, including population densities, of states that have either over 1000 persons per square mile (high density) or under 10 persons per square mile (low density).

The layout of the analysis is to be suitable for a 120-position printer, but can otherwise be in any form you think suitable, provided all titles, headings notations, and comments are clearly shown.

Solution to Problem 5. The information required includes:

1. *General description:* The run reads cards punched as per Fig. 10.7 and produces a printed analysis as shown in Fig. 10.9.

2. *Volume of input:* 51 cards.

PUNCHING INSTRUCTIONS FORM

Department Education				Queries to H.D.Clifton		
Name of source document Appendix 1 of SABDP				Ref. No. of source document —		
Name of card State data				Electro No. 4-658	Color stripe Nil	
Ref	Field name	Card cols.	Picture	Punch left or right	Fill with	Remarks
	Class of card	1	9	-	-	= 1
A	State	2-21	A (20)	Left	Blanks	
B	Area	22-27	9(6)	Right	Zeros	
C	Population	28-32	9(5)	"	"	Expressed in thousands
D	Capital	33-46	A(14)	Left	Blanks	
E	1st Produce code	47-48	99	Right	Zeros	
F	2nd " "	49-50	" "	"	"	Punch from col. 47 onward leaving unoccupied fields blank.
G	3rd " "	51-52	"	"	"	
H	4th " "	53-54	"	"	"	
I	5th " "	55-56	"	"	"	
J	6th " "	57-58	"	"	"	
		59-80	B(22)	-	-	
E	Cereals	47	9	-	-	= 1 or blank
F	Coal	48	9	-	-	"
G	Coffee	49	9	-	-	"
H	Cotton	50	9	-	-	"
I	Dairying	51	9	-	-	"
J	Fish	52	9	-	-	"
K	Fruit	53	9	-	-	"
L	Furs	54	9	-	-	"
M	Hides	55	9	-	-	"
N	Iron	56	9	-	-	"
O	Manufacturing	57	9	-	-	"
P	Meat	58	9	-	-	"
Q	Metals	59	9	-	-	"
R	Minerals	60	9	-	-	"
S	Oil	61	9	-	-	"
T	Sugar	62	9	-	-	"
U	Timber	63	9	-	-	"
V	Tobacco	64	9	-	-	"
		65-80	B(16)	-	-	

First arrangement of produce data

Second arrangement of produce data

Fig. 10.7 Punching instructions forms for Problem 3

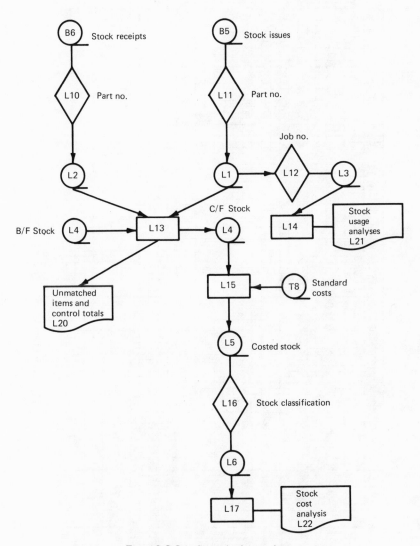

Fig. 10.8 Stock analysis routine

205

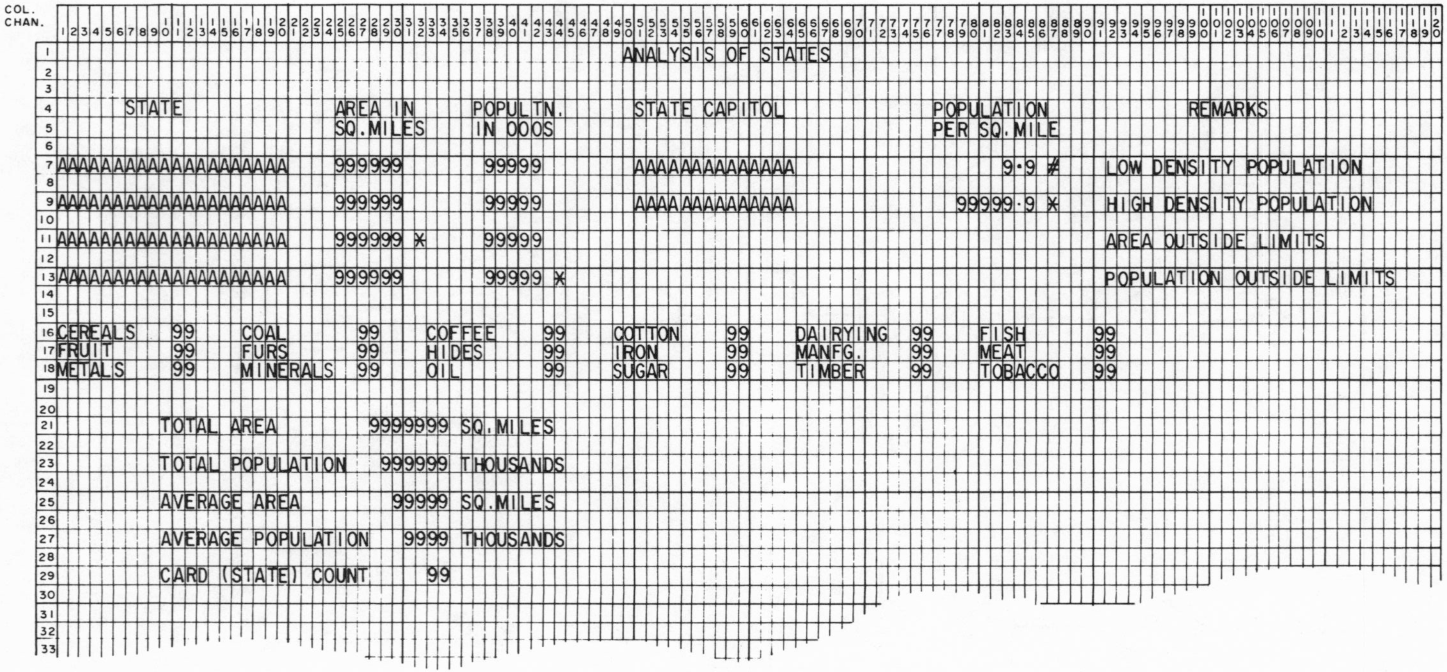
COL.
CHAN.
ANALYSIS OF STATES
STATE
AREA IN
SQ.MILES
POPULTN.
IN 000S
STATE CAPITOL
POPULATION
PER SQ.MILE
REMARKS
AAAAAAAAAAAAAAAAAAAA 999999 99999 AAAAAAAAAAAAAA 9·9 # LOW DENSITY POPULATION
AAAAAAAAAAAAAAAAAAAA 999999 99999 AAAAAAAAAAAAAA 99999·9 * HIGH DENSITY POPULATION
AAAAAAAAAAAAAAAAAAAA 999999 * 99999 AREA OUTSIDE LIMITS
AAAAAAAAAAAAAAAAAAAA 999999 99999 * POPULATION OUTSIDE LIMITS
CEREALS 99 COAL 99 COFFEE 99 COTTON 99 DAIRYING 99 FISH 99
FRUIT 99 FURS 99 HIDES 99 IRON 99 MANFG. 99 MEAT 99
METALS 99 MINERALS 99 OIL 99 SUGAR 99 TIMBER 99 TOBACCO 99
TOTAL AREA 9999999 SQ.MILES
TOTAL POPULATION 999999 THOUSANDS
AVERAGE AREA 99999 SQ.MILES
AVERAGE POPULATION 9999 THOUSANDS
CARD (STATE) COUNT 99

Fig. 10.9 Print layout for Problem 3

3. *Sequence of input:* In alphabetical order of state name (columns 2–21), although this is not consequential.

4. *Meanings of coded fields:* A "one" punched into any of columns 47 through 64 indicates that the state in question yields the produce corresponding to the column as indicated on the card layout (second arrangement) in Fig. 10.7.

5. *Stored data:* The only stored data needed is a table of produce names corresponding to columns 47 through 64; these are needed for inclusion on the printout. The longest produce name occupies 13 bytes, and so the total storage is $18 \times 13 = 234$ bytes.

6. *Processing procedure:* For each card, the following processes apply:
   (a) Feasibility checks and control totals (as described in solutions 7 and 8 below).
   (b) Add area to total.
   (c) Add population to total.
   (d) Examine columns 47 through 64, and when a "one" is encountered, add 1 to the count corresponding to the appropriate produce.
   (e) Calculate population density = (population $\times$ 1000) $\div$ area. Round off the result to one decimal place.
   (f) If population density is under 10.0, print details as per Fig. 10.9, line 7.
   (g) If population density is over 1000.0, print details as per Fig. 10.9, line 9.
   (h) At end of cards, calculate average area = total area $\div$ card count. Round off result to nearest whole number.
   (i) Calculate average population = total population $\div$ card count. Round off result to nearest whole number.
   (j) Print the results of steps (b), (c), (d), (h) and (i) as shown on lines 21, 23, 16-18, 25, and 27, respectively, of Fig. 10.9.
   (k) Print card count from step (a) as per line 29.

7. *Feasibility checks:*
   (a) Check that each area lies within limits 1000 through 160,000, unless state name begins with ALAS, DIST, or TEXA, in which case a limit check is inapplicable; if the check fails, print details as per line 11 and reject the card.
   (b) Check that each population lies between 300 and 20,000; if the check fails, print details as per line 13 and reject the card.

8. *Control totals:* Count cards, omitting those that fail the feasibility checks.

9. *Output layout:* As per Fig. 10.9.

10. *Sizes of totals:* Allowances for number of digits are made as follows:
    Total area = 7
    Total population = 6 (in thousands)
    Produce count = 2
    Population density = 5 plus 1 decimal
    Average area = 5
    Average population = 4

CHAPTER 11

DESIGN OF DATA PROCESSING SYSTEMS—2

11.1 DECISION TABLES

When preparing a computer processing run specification, it is sometimes difficult to describe verbally the certain requirements of the run. As mentioned in Section 10.6, this problem is alleviated by the use of a detailed flowchart, and quite often this is the most satisfactory solution. This is particularly true when the run involves "looping," i.e., repetition of the same series of steps a number of times. Another problem arises when the run includes a number of "decisions" that result in branching of the steps. A few such decisions following one another give rise to a large number of branches, and as a consequence create a complicated and extensive flowchart or verbal description. Only five "yes/no" decisions result in 32 possible branches, and if the decisions are three-way, there are 243 possible branches.

The difficulty of dealing with this abundance of branches is alleviated considerably by the employment of "decision tables." A decision table is a method of specifying concisely the branching rules of a processing run, and is used either instead of or in addition to a detailed flowchart. There is no logical reason against specifying the complete run in the form of a decision table. Quite often, though, this would introduce unnecessary work, as the descriptive form is adequate.

A decision table is set down in the form of a table of figures and words subdivided into four parts (separated by double lines), as shown in Fig. 11.1. Of the four parts, the upper two, "condition stub" and "condition entries," describe the conditions that are to be tested. The two lower parts describe

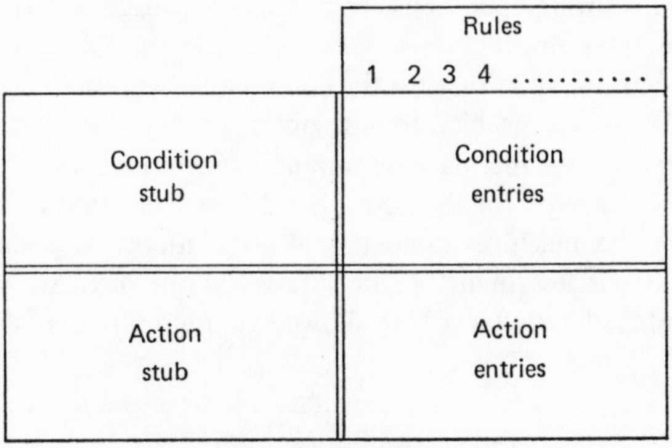

Fig. 11.1 Decision table layout

the actions to be taken, dependent upon the outcome of the tests. A "rule" is in a single vertical column of the table and consists of a set of outcomes of condition tests together with the associated actions. The conditions entries obey "AND" logic; that is to say, if there are two or more condition entries in a rule's column, all conditions must be satisfied before the rule can be applied. Similarly, "AND" logic is applicable to the action entries of a rule, and in some cases the sequence of the actions is also important.

Types of Decision Tables

The entries in a deciison table may be in either limited or extended form; this applies to both conditions and actions. A limited entry is where the stub completely specifies the condition or action; the condition entry then consists of "yes," "no," or a dash (meaning nonpertinent); the action entry consists of a cross if applicable, or a blank otherwise. An extended entry is where the specification is only partially made by the stub and is completed by the entry.

Four arrangements are used:

1. Limited conditions and limited actions.
2. Limited conditions and extended actions.
3. Extended conditions and limited actions.
4. Extended conditions and extended actions.

Examples of these four arrangements, relating to the flowchart in Fig. 11.2 and to the descriptive specification that follows below, are shown in Fig. 11.3.

Suppose we are applying a decision table to a production control application in which it is necessary to decide the interval times to be allowed between manufacturing operations. In ordinary language, the rules to be applied could be stated thus:

"If the operation is the first for the component being made, the interval is zero. Otherwise, if the machine for the operation is the same as that for the previous operation and the machine number is 99 or less, the interval is again zero. When the machine for the operation differs from that for the previous operation and the machine number is 99, the interval is a halfday; if the machine number is less than 99, the interval is one day. All operations on machines numbered 100 upward are allowed an interval of four days."

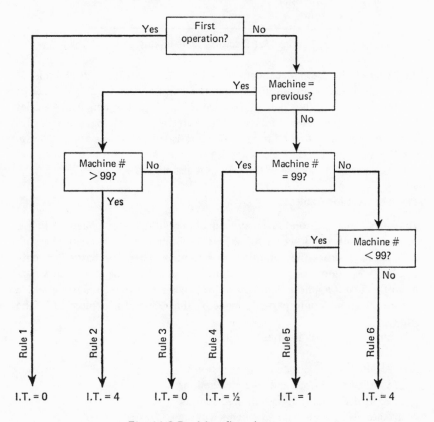

Fig. 11.2 Decision flowchart

Although this statement is accurate, it is not easy to follow; the requirements are more easily understood from Fig. 11.2, or when shown as one of the four alternative arrangements in Fig. 11.3.

(a) Limited condition/limited action decision table

Rule	1	2	3	4	5	6
First operation ?	Y	N	N	N	N	N
Machine = previous?	—	Y	Y	N	N	N
Machine # = 99 ?	—	—	—	Y	—	—
Machine # < 99 ?	—	—	—	—	Y	—
Machine # > 99 ?	—	Y	N	—	—	Y
Interval time = 0	X		X			
Interval time = ½				X		
Interval time = 1					X	
Interval time = 4		X				X

(b) Limited condition/extended action decision table

Rule	1	2	3	4	5	6
First operation ?	Y	N	N	N	N	N
Machine = previous ?	—	Y	Y	N	N	N
Machine # = 99 ?	—	—	—	Y	—	—
Machine # < 99 ?	—	—	—	—	Y	—
Machine # > 99 ?	—	Y	N	—	—	Y
Interval time =	0	4	0	½	1	4

(c) Extended condition/limited action decision table

Rule	1	2	3	4	5	6
Operation	First	Not first	Not first	Not first	Not first	Not first
Machine	—	As previous	As previous	Not as previous	Not as previous	Not as previous
Machine #	—	> 99	≤ 99	= 99	< 99	> 99
Interval time = 0	X		X			
Interval time = ½				X		
Interval time = 1					X	
Interval time = 4		X				X

(d) Extended condition/extended action decision table

Rule	1	2	3	4	5	6
Operation	First	Not first	Not first	Not first	Not first	Not first
Machine	—	As previous	As previous	Not as previous	Not as previous	Not as previous
Machine #	—	> 99	≤ 99	= 99	< 99	> 99
Interval time =	0	4	0	½	1	4

Fig. 11.3 Decision tables

11.2 COMPUTER RUN TIMING

The reasons for estimating the computer run times when designing a data processing system are threefold:

1. The estimated time serves as one basis for comparing the various methods of achieving the desired results.
2. It helps the systems analyst to decide what computer configuration is needed in order to carry out all applications within their processing cycles.
3. The scheduling of computer runs becomes possible, thus giving an estimate of the overall work load on the computer and the data processing operating staff.

In relation to the first reason, there are invariably several methods of achieving the required outputs. Provided there are no overriding objections, they should all be time-estimated as a basis of comparison.

Factors Involved in Run Timing

The estimation of computer run times is a fairly complex subject involving a large number of factors. They can be summarized as:

1. The volume of data to be processed in the run during input, output, and intermediate stages.
2. The computer configuration, especially with regard to the speed of its peripheral units and, to a lesser extent, its processor's cycle time.
3. The mode of operation of the processor, including the amount of overlap of processor and peripheral operations, the degree of multiprogramming that is possible, and the efficiency of the operating system.
4. The types of files, their size, access times, transfer rates, and modes of storage and access.
5. The efficiency of the program, which obviously includes the programmer's skill but is more affected by the programming language and the efficiency of the compiler.

We must bear in mind that at the system design stage, the analyst cannot expect to obtain absolutely accurate timings. This would be possible only after writing the computer program and then carefully aggregating the microseconds for all instructions. Even then the time would be of doubtful accuracy because of program branching, looping, and overlapping. Quite apart from the fact that this task is extremely tedious is that it is unnecessary, for an estimate to within 10 percent is perfectly satisfactory and will not load the computer to its limit, which is preferable. Test runs should enable the timings to be confirmed approximately.

For a given computer run, all or some of the following times contribute to the overall run time. Their relationships are, however, somewhat intricate and dependent upon the particular model and configuration of computer being used.

1. *Peripheral times*: These are, for the most part, the times taken to read punched cards and/or paper tape, and to print the output on the line printer. In general, the times taken by these relatively slow peripheral units are the dominant factors in business applications; a possible exception is the time taken by magnetic tape sorting in certain applications.

2. *Magnetic tape*: Reading, writing, and sorting times; the latter can be very significant in some cases.

3. *Direct access storage times*: Access to records and their transfer to and from core storage (Section 7.8).

4. *Processing time*: Including calculations, index searching, data organization, internal sorting, and internal housekeeping.

5. *Setting-up time*: A wide variety of factors contribute to this, and their significance depends upon the mode of operation of the computer. With multiprogramming controlled by an operating system, the setting-up time effectively disappears, whereas with a small batch processing computer doing a large number of different runs, setup can take a significant proportion of the total time, Included in setting-up time are

Reading into core of the program from magnetic tape, disks, cards, or paper tape.
Loading of peripherals with cards, paper tape, and stationery.
Loading of magnetic tape reels and removable disks.
Typing of messages on the console typewriter, and setting switches.

The time to allow for setting-up a small computer is from zero to 5 minutes per run, depending on the above factors.

6. *Running allowance*: This covers all the extra time consumed by eventualities such as control checks, misreads, error prints, tape rewinds, and dumping. The allowance again depends upon the mode of operation and the complexity of the run; up to 20 percent of the estimated time should be added.

Overall Timing Method. This method involves the estimation of the time for a run by considering the total volume of data to be handled by each of the units of the computer. In this task the manufacturer's timing manual is invaluable, and from it each unit's time can be assessed. The degree of overlapping of activities (simultaneity) of the configuration must also be established; when this occurs, it is generally the longest of the overlapping times that matters. For instance, if three activities such as card reading, magnetic tape reading, and magnetic tape writing occur concurrently during a run, the extent to which these overlap must be found from the manufacturer's literature. Very probably it would be the card-reading time that predominates, the tape operations being absorbed into the time gaps between card operations.

Where the computer operates in multiprogramming mode, it is essential to balance the programs being run concurrently. This is dealt with automatically if an advanced operating system is employed, but otherwise it is the responsibility of the systems analyst to attempt to balance the jobs. This means that, as far as possible, peripheral-bound jobs should be run along with processor-bound jobs so that the computer's full potential will be utilized all the time. This arrangement may, of course, conflict to some extent with the run scheduling (Section 11.3), as dictated by other factors such as source data availability and preparation, post computer jobs, and output deadlines. This principle of run balancing still holds true, however, and should be taken into consideration when preparing the run scheduling.

Having established the computer's ability to overlap operations, the nominal time for each of the computer's peripherals and other units has to be calculated. This is done quite simply, but certain factors should be taken into consideration.

Input Timing Factors. With punched cards, it is not only the number of cards but also; to some extent, the amount of data in them that is of interest. Spread cards containing a large number of items need more data distribution, and this can cause the card reader to fall below full speed, quite apart from the processing of the items. This is not to say that spread cards give a net slower input; the reduction in speed is minimal, whereas the increase in card contents is usually considerable. A typical example is a speed reduction to, say, 80 percent, and an increase in contents to five items per card, giving a net increase in data input speed of four times.

With paper tape input, the main factor to watch for is that the symbols and gaps do not cause a significant increase in reading time. For instance, suppose a transaction actually involves 16 digits, but in punching it into paper tape it is necessary to insert four symbols to identify fields; therefore, a half-inch gap is left between transactions to facilitate visual inspection. The actual amount of tape occupied is $16 + 4 + 5 = 25$ frames per transaction. This is over half as much again as the actual data, resulting in the effective input speed's being at best only two-thirds the maximum.

Magnetic Tape Timing Factors. This pertains to the effective transfer rate, taking into account the block length and interblock gap length.

1. The significance of transfer rate can be seen in Fig. 11.4, which shows the time taken to read 50,000 records of 100 bytes from a 15,000 bytes per second tape drive (IBM 2415-1).

Block length in bytes	Read time in min. with IBG of 0.6 in.
100	32.2
200	18.8
500	10.9
1000	8.2
5000	6.1

Fig. 11.4 Magnetic tape time

2. Whether each reel or tape must be rewound before another is brought into use applies when only one tape drive is available for a file. If two drives are available, the file's tape reels are "queued" on alternate drives, thereby saving rewind time because one reel can be rewound while the other is being read or written.

3. The sorting time taken by the magnetic tape is important and in some routines this dominates the overall time. Tape sorting timing-tables are available from the manufacturers, and these take into account the factors below:

Transfer rate (read/write speed)
Number of decks used in the sort.
Amount of core storage available as buffer areas.
Input/output channels available.
Transfer direction(s); can the tape be read backward?
Block length and interblock gap length.
Number and size of records to be sorted.

Direct Access Storage Device Timing Factors. The hardware aspect factors are the access times and the transfer rates. The former may include the times taken to position read/write heads, verification of head positioning, latency, and the movement of storage media. The transfer rates tend to vary even on the one device, depending on the location of the required data. Great care must be taken when comparing these devices to make sure that all these factors are understood and taken into consideration.

Output Timing Factors. Punched card and paper tape outputs have similar factors to their inputs and the time taken is estimated in a similar way.

Generally, because card and paper tape punches are very slow peripherals, they dominate the timing of any run using them.

Printed output timing involves a number of factors, some of which pertain only to certain types of line printers. These include

1. The number of lines to be printed.
2. The spacing between the lines and the amount of skipping per document. Skipping may occur over the stationery perforations and also in the body of the document.
3. The range of printed characters needed; this applies to certain models of printers that have a variable range.
4. The width of the print train; this in itself does not affect the printing speed, but with a wide print train it may be possible to print two or more documents at the same time (spread printing).

Spread Printing. The technique of spread or side-by-side printing is advantageous in situations where a large number of narrow documents are to be printed and a printer with 120 or more print positions is available. The printing speed is likely to be lowered, but since more printing is done at a time, the net result is a higher output rate. It must be remembered that

1. There must be sufficient core storage available to accommodate the output corresponding to a full set of spread documents because all the output is created prior to starting to print a sheet.

2. Programming is made slightly more complicated, but this can be resolved by suitable subroutines.

3. Some of the print positions are wasted because it is necessary to allow for horizontal gaps between the spread documents.

4. When the number of printed lines varies between documents, the output speed is geared to the document with the highest number of lines in a set. Instead of using the average lines per document as a basis for calculating the printing time, an increase of 20 percent more for two documents per set and 30 percent more for three is better. These figures are obviously very approximate, since an accurate assessment could be arrived at only by knowing the statistical distribution of lines per document.

Shown in Fig. 11.5 are the numbers of spread documents that can be accommodated in various circumstances; the table assumes a horizontal print spacing of ten characters per inch.

Width of print on document, in.	Print positions			
	120	120	160	160
	Interdocument gap, in.			
	½	1	½	1
2	5	4	6	5
3	3	3	4	4
4	2	2	3	3
5	2	2	3	2
6	1	1	2	2
7	1	1	2	2
8 and over	1	1	1	1

Fig. 11.5 Spread documents

Example of Run Timing

The principles behind run timing are best seen from a simple example. This is shown in Fig. 11.6 and consists of a routine of three computer runs carried on a small computer. The computer configuration comprises

> One card reader (peripheral unit) at 900 cards per minute.
> One line printer (peripheral unit) at 600 lines per minute.
> Four magnetic tape drives at 20,800 bytes per second.
> Core storage of 32,000 bytes.

The computer's simultaneity covers the operation of

> One tape drive and one peripheral unit *or*
> Computing and one peripheral unit, *or*
> Four tape drives and computing.

The requirements are:

> RUN 1—Read 20,000 spread cards, write the contents to magnetic tape A1, and at the same time print details from certain cards, amounting to 3000 lines of print. The magnetic tape is written so as to hold 50,000 records of 40 bytes in 400 byte blocks; i.e., each spread card creates an average of 2.5 records, duplication of certain data occurring.
> RUN 2—Sort the 50,000 records to produce tape A2.

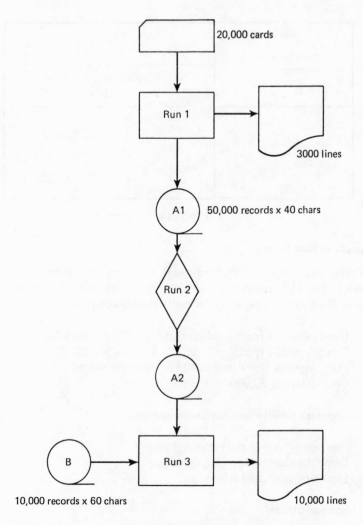

Fig. 11.6 Flowchart

RUN 3—Read the sorted tape A2 and the presorted tape B; the latter holds 10,000 records of 60 bytes in 900 byte blocks. Summarize the records in tape A2, and amalgamate with the records from tape B so as to print 10,000 lines.

Tabulation of results from these three runs is as follows:

	NOMINAL TIME, MIN	NOTES	ALLOWED TIME, MIN
RUN 1			
Read 20,000 cards at 900 cpm*	22.2		22.2
Write 50,000 records	3.1	*a*	–
Print 3000 lines at 600 lpm*	5.0		5.0
Compute	4.5	*b*	–
Setup	4.0		4.0
			31.2
10 percent running allowance		*c*	3.1
		Rounded total	35
RUN 2			
Sort 50,000 records	58.0	*d*	58.0
Setup	3.0		3.0
			61.0
5 percent running allowance		*e*	3.0
		Rounded total	64
RUN 3			
Read 50,000 records (tape A2)	3.1		3.1
Read 10,000 records (tape B)	0.7	*f*	–
Print 10,000 lines at 600 lpm*	16.7		16.7
Compute	3.8	*b*	–
Setup	4.0		4.0
			23.8
5 percent running allowance		*c*	1.2
		Rounded total	25
		Total time for the routine	124

NOTES: *cpm = cards per minute; lpm = lines per minute.
a = overlapped with card reading.
b = overlapped with printing.
c = for card handling and/or stationery adjustments.
d = time derived from manufacturer's manual.
e = for tape handling.
f = overlapped with tape A2 reading.

11.3 RUN SCHEDULING

The scheduling of computer runs is done in outline during the system design stage in order to insure that all the work can be fitted into the available computer time. Later, during the implementation stage, the runs will be rescheduled so as to form a precise timetable of computer operations. Before attempting to create a schedule, each routine will need to have been split into a number of computer runs, and each run to have had its time estimated.

As stated in Section 11.2, if the computer to be employed is capable of operating in multiprogramming mode, several processing runs can be carried out simultaneously. The details that follow do not, in fact, allow for multi-programming, but assume that only one job at a time can be run on the computer. To run-schedule a multiprogramming computer, it is first necessary to find the information appertaining to points 1, 2, and 4–6 in the list below and then to balance groups of jobs for concurrent scheduling. This process is rather more complicated and is not worth planning in great detail at this stage of systems design; in reality, job scheduling will become a tactical exercise later for the chief operator in utilizing the operating system.

In addition to the processing runs, the times taken for data preparation need estimating, and the times of availability of source documents need to be determined. Some of the routines have a deadline time by which the output results must be ready. Failure to meet these deadlines results in delays in dispatching, production, payment of staff, and so on. These deadline routines are given highest priority in the scheduling procedure.

For each routine in the program of work the following information is required before scheduling is done:

1. At what time are the source documents available, and are they in one batch or do they appear in a steady stream?
2. How long does data preparation take?
3. How long does each computer run take?
4. What are the deadline times for final and intermediate results?
5. Is the output extensive enough to justify splitting it into batches for dispatching?
6. What extra work must be done after the computer output is ready and how long will it take? This work includes tasks such as the decollating, bursting and trimming of stationery, the checking and editing of results, the enveloping and dispatching or delivery of the output.

With this information in hand, the high-priority routines are fitted into the schedule, followed in turn by those of decreasing priority. The high-priority routines themselves are best dealt with in order of spare time, those with the least spare time being scheduled first. In this context, spare time is the difference between deadline time and source document availability time less the times for data preparation, computer running, and post-computer work. As an instance, if a batch of source documents is available at 11 A.M. and the deadline is 5 P.M., the punching takes 2 hours, computing 1½ hours, and post-computer work, 1 hour, the routine's spare time is 17 (5 P.M.) − 11 − 2 − 1½ − 1 = 1½ hours.

When a routine has spare time available, it can be maneuvered within its spare time in order to facilitate accommodation of other routines. It is preferable, if possible, not to separate the runs forming a routine but to carry them out in close succession; if separation is unavoidable, it is applied to the lower priority routines first. When positioning a routine within its spare time, the analyst is torn between making it as early as possible—and thereby risking the disruption of other routines if the source documents are late—and leaving it as late as possible, thus risking delayed output results if the computer runs into trouble. A way to avoid this dilemma is to adopt, if possible, a compromise positioning of the routine. This is decided by balancing the probability of documents' being late against the probability of computer failure. Since it is unlikely that either of these probabilities is known precisely, a reasonable plan is to split the spare time fifty-fifty before and after the routine. Once a routine has been scheduled, it may well be necessary to move it within its spare time so that lower priority routines can also be accommodated. An example of this is seen in Fig. 11.7, in which the Wednesday payroll computing has been moved forward so that production planning can be fitted in before its deadline.

Data Processing Scheduling Chart

This is shown in Fig. 11.7 and is based on the computer's switched-on time. This depends upon the number of shifts that will be worked; a fairly safe assumption is initially two shifts amounting to 12 hours per day for five or six days of the week. Out of this time a number of hours is lost to program testing, scheduled maintenance, computer breakdown (sometimes known as "unscheduled maintenance"), reruns due to breakdown, program failure, erroneous input data, and operating mistakes. The manufacturer will specify the time needed for scheduled maintenance and might even venture a suggested allowance for breakdown. Program testing tends to decrease in volume as more work is loaded onto the computer, so that to some degree these two times tend to balance out. It is difficult to allow at all accurately for reruns, but fortunately they tend to diminish with time; initially an allowance of 25 percent of productive time is not unreasonable.

When filling in the computer run-scheduling chart, it is advisable to start with the deadlines and source document availability times for each routine. Thereafter, the routines are dealt with in order of priority, and it is convenient to work from the top of the chart downward. A certain amount of maneuvering is needed to fit in the computing and data preparation; this is facilitated by using a chart with movable indicators. Under each routine the cumulative "bookings" of computer and data preparation departments are shown; these again facilitate the scheduling of the remaining routines. It must be remembered that although the computer works from, say, 0800 to 2000 hours, the data preparation department works scheduled hours, and beyond these times the staff are on overtime rates.

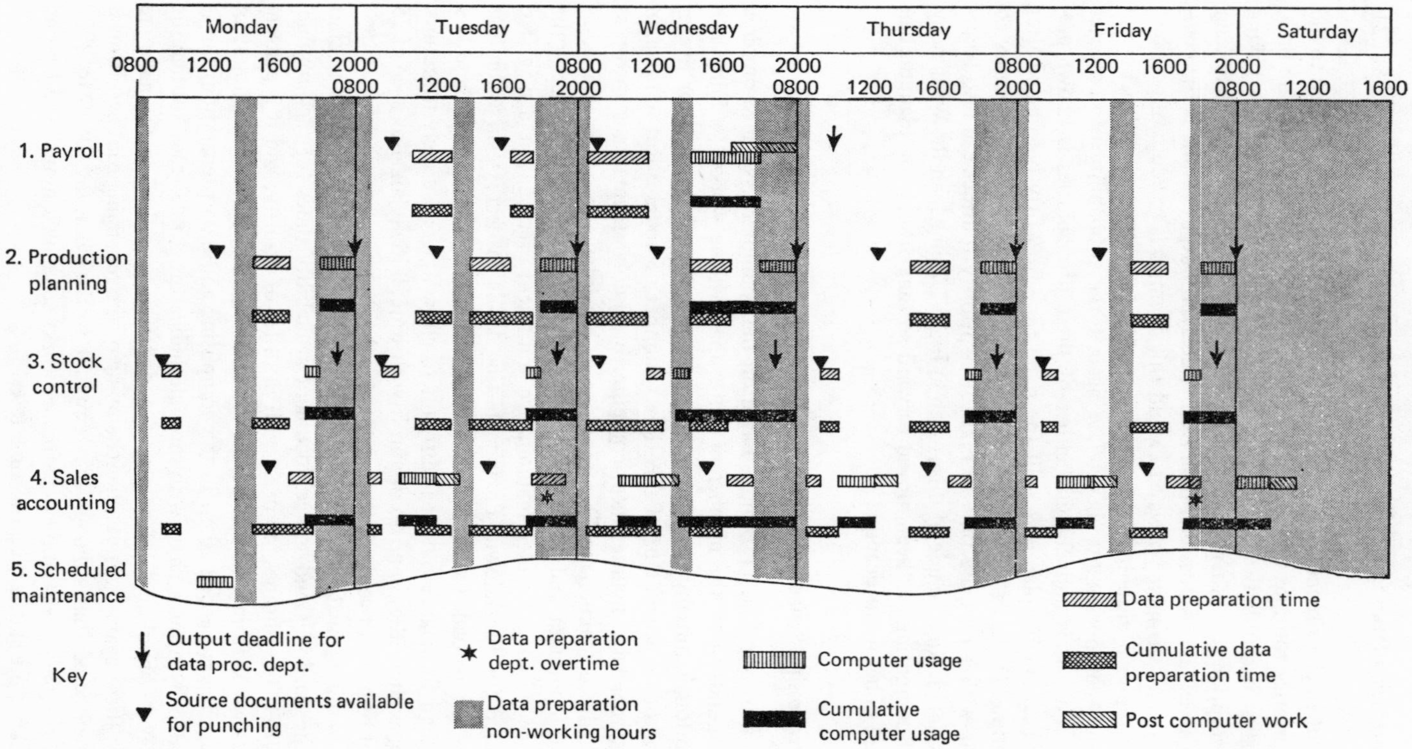

Fig. 11-7 Data processing scheduling chart

Key

→ Output deadline for data proc. dept.

▶ Source documents available for punching

★ Data preparation dept. overtime

Data preparation non-working hours

Data preparation time

Cumulative data preparation time

Post computer work

Computer usage

Cumulative computer usage

1. Payroll
2. Production planning
3. Stock control
4. Sales accounting
5. Scheduled maintenance

Monday   Tuesday   Wednesday   Thursday   Friday   Saturday

0800 1200 1600 2000 0800 1200 1600 2000 0800 1200 1600 2000 0800 1200 1600 2000 0800 1200 1600 2000 0800 1200 1600 2000

11.4 AUDIT REQUIREMENTS

The responsibilities and aims of the auditor are not changed when the routine accounting work is transferred to a computer. His methods of achieving his aims and fulfilling his responsibilities do change, however. The tendency is for his work to involve less checking of figures, leaving more time to investigate the causes of trouble. Much of the data that he has become accustomed to using during auditing is no longer there in the same form. The permanent recording of all transactions no longer occurs, as the nature of computer recording is to make interim data more transient. The auditor can nevertheless obtain details of the system, and although the "audit trail"* in some ways becomes less distinct, it is also less confusing because the auditor is not confounded by a mass of documents and figures.

The systems analyst and the auditor both have a vested interest in insuring the accuracy and the validity of the data processing system. It is therefore the responsibility of the systems analyst to provide the control totals and checks necessary for the auditor to do his job, and it is the duty of the auditor to do his job, and it is the duty of the auditor to adjust his approach to his work so as to dovetail into data processing methods. In order that the auditor can do this with confidence, it is essential that he be trained in computer techniques. Although he cannot be expected to know as much as the analyst or programmer about the details of a particular system or program, an understanding of the methods used is vitally necessary. This means that the auditor must know the meaning of data processing terminology and the purposes of hardware and software.

System Controls

The perpetration of fraud in a data processing system is not impossible, but it is extremely difficult for one person if comprehensive controls are built into the system. A successful fraud would require the connivance of several persons together with a considerable amount of skilled effort on their part. This would be much more the case than in a semimanual system where there is far more opportunity to make unauthorized and undetected alterations to documents during the course of their life.

A more pertinent feature of systems control is the detection of accidental errors. These may be due to a number of different causes, such as:

Omission of input data.
Obsolete input data.
Incorrect master files, incorrectly updated previously or wrong file used.

*Clifton and Lucey, *Accounting and Computer Systems*, Petrocelli Books, New York, 1974, chap. 7.

Undetected program errors, especially after program amendments.
Malfunctions of hardware, especially misreading and misprinting.
Out-of-sequence movements or master files.
Misunderstanding or careless use of operating instructions.

What can the systems analyst do to reduce the chance of accidental errors in the data processing system?

Control Totals. A set of control totals should accompany the data all the way through a routine and, where feasible, through the application. This implies their creation at the source document stage and reproduction during every computer run. Discrepancies must always be reconcilable and, as far as possible, without recourse to complicated manual procedures. The precise steps to be built into a system's control arrangements obviously vary from one routine to another, but the following are generally suitable:

1. A count of all documents sent for punching and, as far as is reasonable, counts of the items thereon. The documents should be batched and accompanied by a batch slip on which are written the counts.

2. Batched totals of quantities and values on documents also included on batch slips. This may involve too much manual work if every field is totaled; the most significant only are therefore included. If no fields of the document are truly cumulative, a "hash" total is formed by adding other fields such as code numbers. Although this has no real meaning, it does help to insure completeness and accuracy.

3. The above-mentioned counts and totals are checked after the data has been punched into cards or tape. With cards, this task can be done off line by using a punched card balancing tabulator; paper tape is not too easily totaled off line except in the case of by-product tape, which can be totaled during its preparation.

4. At the end of each batch in the computer input run, the counts and totals are printed; alternatively they may be stored and printed as a table after reading all the batches. As an additional security measure the control counts and totals themselves may also be input and checked automatically against the figures accumulated by the computer. The details of any discrepancy are then printed and, if desired, the computer is programmed to "lock" so that special operator action has to be carried out before the run can proceed. This leaves less chance of ignoring a discrepancy—deliberately or unintentionally.

5. After each run in a routine, the control counts and totals are again printed, including any new figures created as a result of the processing. The latter are especially important if they relate to data that is used in the subsequent runs.

6. Amendments to, or updating of, a master file should be accompanied by a simple analysis of the master file before and after the run. This includes the numbers of records and any suitable field totals. The analyses are dated and retained as a visible history of the file; they can also be held in the file itself to cover a period of a year or so. *Example of analysis* (control account):

B/F FILE	12,750 records	$33,348 opening balance
INSERTIONS	393 records	$10,200
DELETIONS	258 records	$ 6,314
AMENDMENTS	606 records	$ 3,792 added
C/F FILE	12,885 records	$41,026 closing balance

7. A console log is printed by the console typewriter or written manually. This is preserved for reference purposes so that the auditor can query the reasons for reruns and assure himself that no unauthorized computing has occurred.

8. All computer printed output must be clearly identifiable in terms of its exact particularity. This entails the precise labeling of every sheet of print with its appropriate captions, date, page number, and, in certain instances, line numbers also.

9. The security arrangements for on-line and real-time systems are more exacting and involve stringent control over the use of terminals. Each terminal operator must be identified to and accepted by the computer before being permitted to have access to files. Additionally, real-time involves complex procedures for dealing with breakdowns so that no messages are lost or duplicated and so that audit requirements are satisfied. This is too large a subject to cover in detail here; the reader is advised to consult the literature* for full details. Additional information on the auditing of data processing systems generally is also available.†

Selective File Printing

Every data processing system needs some means of interrogating its files. It is unsatisfactory from not only the auditor's viewpoint but also from that of other staff to have data locked away and inaccessible. In more sophisticated systems file interrogation is an inherent feature, but in the majority of business data processing it is performed by a special run, using a specially written program. As far as is possible, one "selective file print" program should be written as a means of ascertaining what data is held on any file in the system.

---

*Martin, Design of Real-Time Computer Systems, Prentice-Hall, New Jersey, 1967, chap. 35.

† Clifton & Lucey, Accounting and Computer Systems, chap. 9.

This is achieved by the insertion of parameters to describe an individual file before using the selective file print program.

The aims of this program are:

1. To be able to print details of any selected record or group of records on a file.
2. To be able to accumulate and print field totals from any selected group of records, including the entire file.

For interrogating sequential files, either magnetic tape or direct access, this is a fairly straightforward procedure. The computer scans the file sequentially until it comes to the first record within a wanted group. It then prints and/or accumulates details of all following records until the last within the wanted group is reached. Scanning then continues until the next group is reached, and the process is repeated.

Selection from random files is dealt with in one of two ways:

1. Read sequentially into the computer the full set of wanted keys. Access each record in turn by employing the appropriate address generation algorithm, and then print or accumulate its details.

2. Read and store the set of wanted keys, either individually or as the smallest and largest of wanted groups. Scan through the random file in serial access mode, comparing each file key in turn against each of the wanted keys, and then printing or accumulating as necessary.

11.5 CAPABILITIES OF COMPUTER CONFIGURATION

The determination of run times and the subsequent scheduling of the computer's work load are based upon the capabilities of a particular configuration of computer. This may be a hypothetical configuration when the timing and scheduling are first planned, and these tasks should be repeated for several such configurations. The characteristics of these hypothetical computers should be based upon those available with real computers so that when the most suitable has been decided upon, similar actual configurations can be given consideration. This means, of course, that the systems analyst must be cognizant of current hardware and also of available software such as system simulators, which will help in this task.

The power of a business computer is largely determined by four main factors: Its file storage capacity, the speeds of its peripherals and storage devices, language facilities, and the operating system. In deciding at what level these should be chosen, the analyst must look to the future as well as the present. An estimate of future volumes of computer work, based upon a forecast of the company's future activities, may tip the scales in favor of a particular model and configuration. In a similar way it may be discovered that the times taken for most of the runs hinge upon the speed of one peripheral; this is an obvious case for considering more or faster peripherals for this type.

Compatibility

One facet of computer compatibility relates to the ability of a particular line of computers to be enhanced by attachment of further and/or faster peripheral units, storage devices, core store, etc. This is termed "standard interface," and means the ability to expand a computer as required. All modern ranges of computers have this facility so that a user can initially purchase a small computer and expand it later by adding units as and when they are needed.

Another facet of compatibility is related to magnetic tape, some types of which are now compatible between different computer ranges. This means that, if desired, magnetic tape can be freely exchanged between computers.

A third aspect of compatibility is with regard to the programming languages available with the computer. In cases where existing computer work has to be transferred to a new computer, language compatibility is invaluable. All computers now have compilers capable of accepting COBOL and FORTRAN, although slight differences do exist.

Core Storage

Although there is no formulative relationship between the amount of core storage and the general size of the computer, the two tend to increase together. When estimating the required amount of core, the systems analyst must insure that there is sufficient to accommodate not only the user's own programs but also the software supplied by the manufacturer. This consists mainly of the operating system and application packages.

Allowance should also be made for core storage to hold indexes, tables, buffer areas, etc.

Magnetic Tape

A glance at the manufacturers' manuals shows that the sorting speed on magnetic tape is related to both the number of tape drives used and the transfer rate of the tape. The decrease in time taken is most marked between three and four drives, the extra drive reducing the time by roughly 25 percent, as can be seen from the manufacturers' manuals. With additional drives, this decrease is less marked; therefore, unless a tremendous amount of sorting has to be done, four drives are sufficient. Sorting speed is more or less proportional to the tape transfer rate so that the problem of sorting is one of balancing the number of decks against the tape transfer rate selected in order to obtain a given sorting time.

Most computer runs other than sorting use three or four decks, with a few needing five or six. Unless the minimization of run time is of paramount importance, serious consideration ought to be given to splitting the five-drive runs into two simpler runs. Often, a convenient way of doing this is to combine input tapes by merging in a prerun, or to reduce output tapes in a run by using one output tape and separating the data in a post-run. A most unsatisfactory state of affairs exists when only one or two runs in the data processing system require five drives and the remainder require only four or less. It is then well worth the analyst's efforts to reduce the maximum requirement to four, bearing in mind that in practice one drive may be permanently occupied by program tapes. If it is not possible to reduce the number of tape drives required, consideration ought to be given to drives of a lower transfer rate in order to minimize the overall cost.

Disks and Other Direct Access Storage Devices

The three main criteria to be applied to the selection of direct access storage devices are storage capacity, access time, and transfer rate. Storage capacity is closely connected with the exchangeability of the storage media, and therefore it is the on-line storage requirement that is the determining factor. When dealing with the latter two criteria, the degree of their importance depends on the modes of file access and storage, and on the files' activities. The access time and transfer rate must be consistent with the requirements of the system, but generally, provided the files are properly organized, they do not significantly affect computer run times of the more conventional business applications.

Another factor to consider when selecting storage capacity is the relatively high cost of disk packs and other direct access storage media, which can be justified only by full utilization of the capacity. This is achieved by arranging files so as to fill removable disk packs as much as possible, at the same time remembering to allow for expansion and overflow. The employment of nonremovable disks is best limited to applications in which the files may be retained permanently on the disks. The amount of data held on nonremovable disks is generally so large that it is not a viable proposition to load and unload the disks except from high-speed magnetic tape, and this combination of disks and tapes results in an expensive configuration.

11.6 EXERCISES

Problem 1. Decision Tables

Create (a) a limited-condition/limited-action decision table, and (b) an extended-condition/extended-action decision table to convey the following information: Discounts are allowed to customers based upon three factors: type of customer, wholesale, retail, or individual; long-standing or new; large orders ($1000 and up), medium orders ($100-999), or small orders (below $100). Long-standing customers except individuals receive discounts of 25 percent for large orders, 20 percent for medium, and 15 percent for small. The corresponding discounts for individual customers are 10, 7½, and zero percent, regardless of standing. New wholesalers get 20 percent on large orders; otherwise, 5 percent. New retailers get 10 and 5 percent correspondingly.

Solution to Problem 1. (a) As in Fig. 11.8, and (b) as in Fig. 11.9.

Problem 2. Run Timing

A computer consists of a card reader operating at 300 cards per minute, a line printer operating at 250 lines per minute, and 4 magnetic tape drives operating at 15,000 bytes per second. Assume that these devices cannot operate simultaneously and that the tape times are proportional to the number and length of the records in Fig. 11.4. Estimate the approximate times taken for the runs listed below. Setup and computing times are to be ignored; running allowances are 3 percent for each device used.

(a) Read 10,000 cards and write the contents to tape in the form of 1000 blocks of 500 bytes, holding 10 records in each block.

(b) Read 20,000 cards to be matched against a tape holding 50,000 records of 200 bytes in 1000-byte blocks, and print 1000 lines.

(c) Read two tapes, A and B. Print 2000 documents averaging 20 lines and taking an average of a half-second per document for spacing and throwing. Tape A holds 30,000 records of 100 bytes in 500-byte blocks; tape B holds 20,000 records of 40 bytes in 200-byte blocks.

(d) Read 15,000 spread cards, merge with a tape of 20,000 records of 250 bytes in 1000-byte blocks, and write to tape in the form of 60,000 records of 25 bytes in 200-byte blocks. Ten percent of the cards have exception prints of one line, and 2 percent involve printouts of ten lines plus a document movement time of a quarter-second.

Solution to Problem 2. Approximate run times in minutes are:

(a) Read 10,000 cards at 300 cards/minute   = 33.3

     Write 10,000 records of 50 bytes = $10.9 \times 1/5 \times 1/2$ = 1.1

                                                          34.4

     Running allowance = $2 \times 3\% = 6\%$ = 2.1

                                 Rounded total   = 37

	1	2	3	4	5	6	7	8	9	10
Wholesale?	–	–	–	–	–	–	Y	Y	–	–
Retail?	–	–	–	–	–	–	–	–	Y	Y
Individual?	N	N	N	Y	Y	Y	–	–	–	–
Long-standing?	Y	Y	Y	–	–	–	N	N	N	N
Large order?	Y	–	N	Y	–	N	Y	N	Y	N
Medium order?	–	Y	N	–	Y	N	–	–	–	–
Discount = 25%	X									
Discount = 20%		X					X			
Discount = 15%			X							
Discount = 10%				X					X	
Discount = 7½%					X					
Discount = 5%								X		X
Discount = 0						X				

Fig. 11.8 Limited-condition/limited-action decision table

Customer type	–	–	–	I	I	I	W	W	R	R
Standing	E	E	E	–	–	–	N	N	N	N
Order size	L	M	S	L	M	S	L	M or S	L	M or S
Discount =	25	20	15	10	7½	0	20	5	10	5

I = individual, W = wholesale, R = retail

E = long-standing, N = new

L = large, M = medium, S = small

Fig. 11.9 Extended-condition/extended-action decision table

(b) Read 20,000 cards at 300 cards/minute   =   66.7
    Read 50,000 records of 200 bytes = 8.2 × 2   =   16.4
    Print 1000 lines at 250 lines/minute   =   4.0
                                          87.1
    Running allowance = 3 × 3% = 9%   =   7.8
                         Rounded total   =   95

(c) Read tape A:

$\qquad$ 30,000 records of 100 bytes = $10.2 \times \frac{3}{5}$ $\qquad$ = 6.5

Read tape B:

$\qquad$ 20,000 records of 40 bytes = $18.8 \times \frac{2}{5} \times \frac{4}{10}$ $\qquad$ = 3.0

Print 40,000 lines at 250 lines/minute $\qquad$ = 160.0

Transport 2000 documents at $\frac{1}{2}$ second $\qquad$ = 16.7

$\qquad\qquad\qquad\qquad\qquad\qquad\qquad\qquad\qquad\qquad$ 186.2

Running allowance = $3 \times 3\% = 9\%$ $\qquad$ = 16.8

$\qquad\qquad\qquad\qquad\qquad\qquad$ *Total* $\quad$ = 203

(d) Read 15,000 cards at 300 cards/minute $\qquad$ = 50.0

Read 20,000 records of 250 bytes = $8.2 \times \frac{2}{5} \times 2.5$ $\quad$ = 8.2

Write 60,000 records of 25 bytes = $18.8 \times \frac{6}{5} \times \frac{1}{4}$ $\quad$ = 5.6

Print 1500 lines at 250 lines/minute $\qquad$ = 6.0

Print 300 lines at 250 lines/minute $\qquad$ = 1.2

Transport 300 documents at $\frac{1}{4}$ second $\qquad$ = 1.3

$\qquad\qquad\qquad\qquad\qquad\qquad\qquad\qquad\qquad\qquad$ 72.3

Running allowance = $4 \times 3\% = 12\%$ $\qquad$ = 8.7

$\qquad\qquad\qquad\qquad\qquad$ *Rounded total* $\quad$ = 81

Problem 3. Run Scheduling

During a systems design phase, the work outlined below is to be scheduled on a small computer that is capable of carrying out only one job at a time. The computer is to be employed continuously from 9 A.M. to 8 P.M. Monday through Friday, and the data preparation section works on those days from 9 A.M. to 1 P.M. and 2 P.M. to 5 P.M. Schedule the jobs so as to involve minimum overtime and yet enable all jobs to meet their deadlines. The jobs are shown in order of priority.

(a) *Order processing*: Customers' orders arrive daily by post in time to start data preparation at 10 A.M. This takes an average of 3 hours and is to be followed by the computer run, averaging 4 hours. Computer processing must be completed in time to start dispatching the goods early next morning.

(b) *Payroll*: The payroll source data is available by midday Tuesday. The computer output must reach the pay office by 9 A.M. Thursday. Data preparation takes 4 hours and computer processing 2 hours.

(c) *Stock vetting*: A daily stock report is to be prepared immediately after order processing. The stock receipts will be available by 9 A.M. and these, together with the results from order processing, will form the source data for this run of 2 hours; data preparation takes 1 hour. The stock report is to be ready by 9 A.M. next day.

(d) *Purchase accounting*: Documentation is due to arrive from the purchasing department around 5 P.M. each day; data preparation occupies 1 hour and computer work 2 hours per day. There is no absolute deadline, but the work should be completed as early as possible in order to send off the purchase orders quickly.

(e) *Sales accounting*: This is carried out weekly, based upon the results from order processing. The weekly data preparation time averages 4 hours and the computer run time 6 hours. There is little urgency, and so this work can be fitted in between other jobs; runs should not be less than 1 hour each, however.

Solution to Problem 3. As can be seen from Fig. 11.10, it is possible to fit these five jobs into the weekly hours—but only just! There is the necessity for 3 hours overtime on the part of the data preparation section, and the purchase accounting work on the computer is somewhat fragmented. In Fig. 11.10, the small letters near each piece of data preparation and computer processing signify the day of origin of the data.

In reality, a situation resulting in such a heavy load on the computer calls for an extension of its shift hours in order to allow for breakdown, maintenance, and reruns. It is also possible that Saturday or Sunday working might be needed, provided this is acceptable to the staff—probably not!

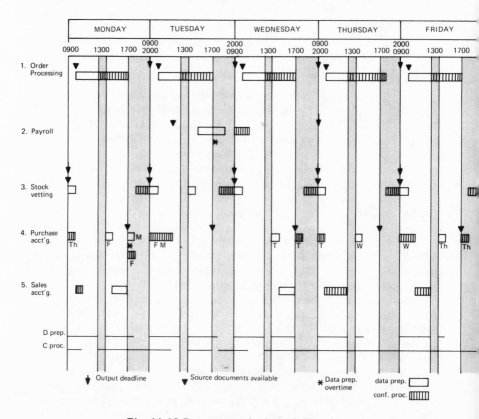

Fig. 11.10 Data processing scheduling chart

CHAPTER 12

APPRAISAL OF DATA PROCESSING SYSTEMS

12.1 COMPARISON WITH EXISTING SYSTEM

When the data processing system has been designed, the systems analyst should evaluate it in comparison with the system that it is intended to replace. This comparison should cover not only the advantages of the data processing system over the existing system, but also an honest assessment of its disadvantages. By making himself aware of these advantages and disadvantages, the systems analyst is able to forestall criticism of the new system when he presents it to the management and staff.

The comparable aspects of the data processing system and the existing system are financial, speed of throughput, and management information and control (Section 12.2.). These aspects should be given consideration when making the comparison both before and after implementation (Chapter 13). When all is said and done, appraisal before implementation is entirely theoretical, since no actual measurements of performance are as yet possible. Nevertheless, it is still a worthwhile and very necessary exercise, for without it the systems analyst can have no idea as to the efficacy of his system.

Financial Comparison

The existing system will have been cost-estimated previously (Section 4.8). Now the costs of the proposed data processing system are to be estimated, bearing in mind that they will not be constant from year to year. An exact mathematical comparison of the two sets of annual costs involves the use of "discounted cash flow" techniques (Section 9.5).

The financial factors that enter into the installation of a data processing system are equipment, environment, materials, staff, and training costs.

Equipment Costs. Quite apart from the range of equipment involved in the system, a prime consideration is the method of purchase. The merits of outright purchase, rental, leasing, or other financial arrangements ought to be studied by the financial executives of the organization, guided by the analyst's advice on the hardware aspects (Section 9.5). The costs of capital equipment include not only those for the computer and its on-line peripherals, but also items such as:

Terminal equipment and data transmission equipment.
Card punch/verifiers.
Paper tape punch/verifiers.
Document readers.
Stationery handling equipment (bursters, guillotines, etc.).
Ancillary punched card machines (sorters, reproducers, balancers).
Extra reels of magnetic tape.
Extra removable disks, magnetic cards, etc.
Trays, trolleys, racks and cabinets for cards, paper tape, magnetic tape, and disks.
Additional furniture and clocks.
Bench, cabinets, and furniture for the maintenance engineer (some manufacturers include this equipment as part of the contract).

Environment Costs. These include all factors connected with the preparation of a computer room. Sometimes this involves the erection of a new building or large-scale constructional changes to an existing site. In any case, even if an existing room is to be used, some or all of the following will be needed:

Removal and resiting of existing department(s)
Air conditioning plant and associated control gear (capital and maintenance costs).
Lighting.
Acoustic and thermal insulation.
Security arrangements.
Floor, wall, and ceiling preparation.
Fire prevention equipment.
Other extra accommodations such as data processing offices, store rooms, ancillary machine room, punch room, maintenance engineer's room, air conditioning plant room, space for water tanks.
Any special arrangements for moving equipment in and out of the building.

Stabilized electric supply and standby power equipment.
Intercom and telephone systems.

Costs of Materials. These are fairly high at first and continue throughout the life of the system; they include:

OCR and OMR documents.
Punched cards.
Paper tape.
Printed stationery and ribbons.
Programming stationery.
Books, manuals, and other literature and stationery.
Air conditioning filters and materials.

Staff Costs. These include all members of the data processing department and also the staff retained in other departments who are connected with the data processing system and who were included in the existing system costs. The data processing department staff consists of:

The manager(s).
Systems analysts.
Programmers.
Computer operators.
Punch operators.
Control staff (tape librarian, control clerks).
A handyman (primarily for moving materials).
Consultant's fees must also be taken into account.

The computer maintenance engineer is usually employed by the manufacturer. When a large number of on-line terminal operators is involved, it is best to estimate their costs on a separate basis, since these may be very considerable. It is generally the situation, however, that these people are existing employees put to a new role and therefore that there is not necessarily any significant net increase in their cost.

Training Costs. These are high initially and continue to a small degree, dependent largely upon the rate of staff turnover. The initial training costs are made up of:

Programming course fees.
Systems analysis course fees.
Management course fees.
Accommodation and traveling expenses of trainees.

Terminal operator training and subsequent computer training mode facilities.

The preceding costs are subject to being reduced by any "free" courses allocated by the manufacturer in relation to the cost of the computer. Also under this category can be included the computer rental charges (if any) and staff expenses associated with program testing prior to delivery of the user's own computer (Section 13.1).

Development and Operating Costs. These costs are sometimes not entirely separable and continue at different levels throughout the life of the data processing system.

Program-testing charges prior to installation of own computer.
Contract programming fees.
Purchase of software, including application packages.
File conversion costs.
Changeover costs such as parallel running.
Rent, heating, cleaning, and electricity costs.
Standby arrangement charges on bureau or other user's computer.
Rentals of modems and transmission lines.
Insurance premiums.

Comparison of Speeds of Throughput

"Time is money"—but this is not relevant if it is someone else's time. When comparing the throughput speeds of the data processing and existing systems, the significant point is: What benefits are accrued from the reduced throughput time?

Every endeavor should be made by the systems analyst to quantify benefits, in dollars if at all possible. Intangible benefits are always suspect, and should not therefore be used as a blanket reason for adopting a data processing system. Improved speed of throughput is meaningful only if quantified in relation to the cost of the improvement. It is up to the systems analyst to attempt to do this wherever possible.

The possible benefits arising from reduced throughput time are:

1. Reduced waiting time in the factory; this benefit is obtained by faster replanning when changes occur.
2. Quicker detection and reporting of situations that require human action.
3. More up-to-date management reports.
4. Better creditor payment control. The faster throughput of these payments means that they can be delayed until the most propitious moment.

5. Earlier collection of debts by more rapid determination of the amounts due.
6. Reduced storage space for stock-in-hand, materials etc., since more rapid reordering means that lower stock levels need to be held.

Cost Comparison Table

In order to demonstrate the comparative costs of the existing system with a proposed data processing system, it is useful to draw up a cost comparison table. Examples of this are shown in Figs. 12.1A and 12.1B, and cover a period of five years from the start of the data processing system. There is no absolute reason for adopting a five-year period, and other periods are equally valid. The potential savings are derived from the staff utilization table (Fig. 4.6) and the departmental costs table (Fig. 4.7). The cumulative net savings should, of course, be assessed by utilizing investment appraisal techniques such as are described briefly in Section 9.5.

It is, of course, extremely difficult to forecast the rise in costs as far ahead as five years. There is little to be done about this except to assume that all costs—existing or data processing—will rise by roughly the same proportion.

Any national or state grants or allowances are offset against the data processing costs, and these must be studied by the company's financial accountant and then applied to the cost comparison table.

12.2 MANAGEMENT INFORMATION AND CONTROL

The paramount consideration with regard to the management information provided by the data processing system is *quality*. This means its accuracy to the degree necessary, its topicality, its conciseness, and its comprehensiveness. The analyst endeavors to design a system that provides all these characteristics in its output of management information. In so doing he finds it necessary to strike a balance between conciseness and comprehensiveness. How can these two apparently conflicting aims be reconciled? He must not present management with a mass of detail such as is only too easy to produce with a computer. On the other hand, whatever information is prepared must be based on all the facts of the situation.

A solution to this problem lies in the provision only of information that leads to action or decision on the part of management; this information is, however, based upon the full complement of relevant data. This principle is termed "management by exception," and the skill in design lies in deciding what is exceptional. Management will have provided some indication of their information requirements when they were interviewed during the investigation, but these requirements may disappear as a result of automatic action by

Item	Type of cost	Financial year				
		1974-5	1975-6	1976-7	1977-8	1978-9
Computer	R	83000	83000	83000	83000	83000
Punches	P	13000	2600			
Punches maintce.	R	600	750	750	750	750
Tape reels	P	2150	600		250	250
Card equipment	P	500	150	150		
Equipment totals		101970	88000	84250	84350	84350
Air conditioning	P	18000				
Air conditioning maintenance	R	350	350	350	350	350
Environment totals		22100	500	500	500	500
Punched cards	R	3350	1900	1900	1900	1900
Stationery	R	1100	250	250	250	250
Materials totals		4800	2300	2300	2200	2200
D.P. mangt. salaries	R	8400	8400	8400	8400	8400
Systems salaries	R	11800	11800	11800	11800	11800
Progrs. salaries	R	16800	21600	21600	21600	21600
Salaries and o/h totals		44400	50400	52800	52800	52800
Training	R	2900	1000	700	700	700
Misc. costs totals		4800	2600	2400	2400	2400
Grand totals		178270	143800	142250	142250	142250

P = Outright purchase (capital) cost
R = Rental or regular cost

Fig. 12.1a Cost comparison table (figures shown do not purport to be realistic).

238

Dept.	Routine	1974-5	1975-6	1976-7	1977-8	1978-9
Stores ctl.	Filing and sorting	16800	35688	35688	35688	35688
	Ledgers		12000	31918	31918	31918
	Stock checkg.			13200	13200	13200
	Purchase reqs.		6000	15600	18000	18000
	Works orders			12000	12000	12000
	Other work	960	2640	4320	4800	4800
Stores ctl. office totals		17760	56328	112726	115606	115606
Purchasing	Ledgers		8640	13920	13920	13920
	Typing		5880	11520	11520	11520
Purchasing totals			21600	31200	54000	54000
Production ctl. totals				13200	28800	43200
Wages dept. totals		14400	24000	24000	24000	24000
Sales invoicing totals			9600	19200	26400	26400
Grand totals (savings		32160	122400	288000	336000	355200
Less grand totals (costs)		178270	143800	142250	142250	142250
		—146110	—21400	145750	193750	212950
Accumulative net savings		—146110	—167510	—21760	171990	384940

Fig. 12.1b Cost comparison table.

239

the data processing system. It is therefore prudent to make further inquiries of management during the design stage. At this time they can be told the general line of approach of the system so as to guide them in redefining or reaffirming their information requirements. Every piece of information has its value, although this value is not easily measured in financial terms. Nevertheless, the systems analyst must make some evaluation of required information so that this can be compared with the cost of producing it. It is obviously unsatisfactory to allow a manager's whim to instigate the adoption of an expensive piece of hardware or a time-consuming computer routine.

The same principle applies to the degree of control that is contemplated; the cost of any improvements in this respect must be commensurate with their value to the company. It is, for instance, pointless to introduce a special checking routine that costs the company $2500 per annum, when the saving made through the checking is known never to exceed $2000 per annum.

Management Information Reports. These are required for four main reasons:

1. *Action*: Often associated with unforeseen circumstances that call for emergency action, this normally involves middle management rather than top management.
2. *Assurance*: That operations within a department or factory are proceeding, within accepted limits, according to plan.
3. *Tactical planning*: The comparatively short-term disposition of the manager's resources, such as machine shop loading for the following week.
4. *Strategic planning*: The longer term planning of future markets and products, and deployment of financial resources.

Management Information Considerations. In designing a management information system,* the systems analyst will always be torn between providing everything that the manager requires by way of information and keeping the system straightforward and within economic limits. Many sophisticated computer-based management information systems have been suggested, but a much lesser number have been put into practical use. Management information appears in general to be disproportionately expensive as compared with other, more mundane, computer output. This is even more so when the required information is changeable from day to day, and the systems designer is inevitably forced to prune the information asked for. Unfortunately, pruning is not easy if he is to be sure that nothing vital is lost and also that trivia do not contribute to high cost.

*Carr, "A survey of management information systems literature," *Computer Bulletin*, Vol. 15, No. 6 (June 1971), British Computer Society.

There are relatively few business situations that justify the employment of conversational mode purely for management's benefit. Conversational computing can provide immediate information of a wide variety and flexibility, but only at high cost in hardware and at tremendous effort in planning and programming. Essentially, the systems analyst must obtain answers to the following questions before committing himself to the design of a management information system involving substantial costs over and above those already incurred for basic data processing.

1. Does the manager who asks for information appreciate its cost?
2. To what extent is his requirement of a temporary nature?
3. Are the parameters of the information likely to change? If so, can these changes be predetermined?
4. How much do each manager's information needs duplicate other requests, not only in content but also in degree of comprehensiveness? Can such information be generalized to serve a wider management audience?
5. What is the time factor involved; If immediate, is there sufficient justification to warrant a rapid response, conversational mode, or on-line system?
6. If the management information requirements are obscure, what additional data should be held on files to satisfy them when they become clear?

Although it is improbable that all the preceding questions can be answered unequivocally, posing them in itself stimulates managers into analyzing their demands more critically—this can do nothing but good!

12.3 DATA PROCESSING FACILITIES

In Chapter 11, the problem of determining the computer's configuration was discussed. It must not be assumed, however, that the purchase of the user's own computer is necessarily the best means of implementing data processing in all situations. There are a number of alternatives to this, and they should all be given due consideration. The alternatives to outright purchase of the computer are renting, leasing, installment plan, or other financial arrangement agreed upon with the supplier. Each of these has its advantages with regard to taxation, depreciation, and government grants and allowances. From the systems design and operating aspects, all three arrangements are identical, since they all involve placing the computer on the user's own premises.

The main alternative facilities to this are batch processing service bureaus, time hire, on-line time sharing, and shared computer.

Batch Processing Service Bureaus

When considering the possible adoption of bureau service, it is vitally important to assess the capabilities of a bureau before allowing it to undertake any work of great extent or immediacy. This problem is not easily dismissed, since there is no direct ratio between a bureau's cost quotation and its degree of efficiency.

What should the systems analyst do, then, if he requires bureau service?

1. Obtain quotations from several bureaus, all based on identical and precise specifications of the work to be carried out. No points should be left in doubt with regard to (a) the exact results required, (b) the source data to be supplied by the customer, and (c) the turnaround times of the operation. The quotations should be broken down into the separate charges for programming, data preparation, file creation, regular processing, and other work.

2. Insure that the bureau staff appreciate the amount of preliminary work involved in creating and checking the files, in vetting basic input data, and in handling the output, if this is appreciable.

3. Study the contractual agreement proposed by the bureau so that there is no doubt as to who is responsible for what.

4. Check up on the bureau's efficiency by (a) inquiring about the standby arrangements for coping with prolonged computer breakdown, and by (b) investigating the bureau staff's competence in systems design and programming. These points are rather difficult to establish, and it is therefore wise to discuss the bureau's efficiency with its existing customers who are having similar work carried out.

5. Ascertain the eventual ownership of the programs and documentation associated with the work in hand. There is always the possibility that the bureau work will latter be transferred onto the customer's own computer, and so program ownership may be consequential.

6. Establish precisely who will be doing each stage of the overall work. Some bureau firms offer a "turnkey" operation, so called because the complete process—from investigation of the customer's requirements through to job running—is supervised by the bureau or its agents. Others expect to act merely as operational agencies, all design and programming being carried out by the customer. Naturally, the respective charges reflect the degree of involvement of the bureau.

7. Owing to the high financial failure rate of service bureaus, it is advisable to investigate their financial stability.

Time Hire

The user hires a computer for a certain length of time at prebooked regular or irregular intervals. At best, this is equivalent to the user's having a computer of his own for a restricted time. At its worst, it is a tedious and time-wasting method of obtaining the desired results. The degree of efficiency obtainable with this facility depends upon the hired computer's configuration, its distance from the user's premises, and the proficiency of usage in relation to the coordination of computer runs and data preparation.

Unexpected costs are apt to arise if events do not go according to plan, and so it is advisable to study carefully the viability of time-hire before committing the data processing work to this method. This aspect is particularly important if computer time is to be hired from another user whose computer is currently underutilized. At a subsequent date, this underutilization may be taken up by the user's overloads and reruns, thus leaving the time-hirer in difficulties.

On-line Time Sharing

With this facility the user shares the computer simultaneously with a number of other users. This is accomplished by the user's premises containing a direct link to the computer in the form of a terminal, data being transmitted to and from the computer over communication lines. The lines are connected to the computer via a line control unit; this, together with the computer's multi-programming ability, enables it to deal with each terminal in rapid succession. Thus, because the computer's response is immediate, the user is not made aware that he is sharing the computer simultaneously with others. Each terminal is equipped to meet the user's needs by the installation of units selected from a variety of peripheral and special equipment.

An important feature of an on-line time-sharing service that needs to be investigated by the prospective user is its existing and intended loading. The prospective user must assure himself that the computer load already committed or planned will not be so high as to cause excessive delays during processing runs. It should also be remembered that an individual company offering on-line time-sharing service might be geared to a certain type of work, and consequently the level of service might not be so high for other types of work.

Although the actual processing costs of this facility are usually quite modest, this might be less true for the data transmission and storage costs involved. The data transmission cost is generally based on a number of factors, such as the distance between the terminal and the computer, the times of day of transmitting, the speed of transmission, and of course the volume of data. With regard to the storage of data, it is obviously costly to pay for on-line storage of large files if these are to be accessed only infrequently. This situation can easily come about gradually and may remain unnoticed, owing to the slow but steady growth of files or their gradual obsolescence.

The various on-line time-sharing service companies offer a wide range of services, including application packages and conversational mode. It is important that the prospective user of this system, if regular accounting work is intended to be done, should enter into the safest contract for his real needs. In addition the service company must demonstrate its ability to provide an economical and continuing service.

Provided these conditions can be met, on-line time sharing is the answer to the data processing problems of a large number of small firms and of the special application user.

Shared Computer

This is not to be confused with time sharing; a shared computer merely means that several users join in purchasing a computer and then apportion its working time among them. The users are often members of the same group or association. The computer may be in a data processing department staffed by "group" employees, or it may be used by a co-owner who has his own data processing staff for programming and operating. These two arrangements are in effect the same as bureau service and time-hire, respectively.

With computer sharing, care must be taken in planning and scheduling the jobs so as to insure that all users can process their requirements punctually. With acceptance of this constraint, the result of sharing a larger computer is normally more economical and efficient than if each user firm had its own smaller machine. There is no real comparison between the capabilities of a computer with large-scale storage facilities and the alternative of a number of visible record or minicomputers controlled essentially by hand.

Factors in Choice of Facility

Obviously, many factors can enter into a particular company's decision regarding the choice of a facility and its application. Some of the factors relate purely to the company's circumstances at one point in time or to its long-term plans. However, three main general factors control the choice: cost, staff involvement, and file accessibility.

Cost. Cost is generally of prime importance when making a choice, but the systems analyst must be certain that he is comparing like with like. It is, for instance, unlikely that a bureau service can cover an integrated or total system. As far as most users are concerned, it functions best when employed for a few clearcut applications. Other more integrated facilities are more comparable with each other, since most all can achieve the same results.

When comparing the quoted or calculated costs of the work, it must be remembered that allowance is necessary for contingencies such as reruns. In the case of time-hire, and possibly with a shared computer, additional costs are introduced by the transportation of input media, output results, and the operating staff itself.

Staff Involvement. The "odd man out" may again be the service bureau. With all other facilities the user's own staff will be deeply involved in planning and operating the system. In the case of bureau service, this may not necessarily be so. If it is the intention to employ a bureau to implement an isolated and straightforward application, there is no need for deep involvement of the user's staff in general. Nevertheless, at least one member should be made responsible for maintaining close contact with the bureau, and for keeping himself informed about the work in hand. This may sound obvious, but it is a fact of life that some users gradually lose real contact with the bureau; consequently, with the passing of time, the results they receive become almost meaningless to them.

If the service bureau work is at all comprehensive or complete, it is essential that the user's systems analysts are effectively in control of the system. They should be of the same caliber as for internal computer usage, since they are equally responsible for designing the system. It is both unwise and unreasonable to expect the bureau's staff to do detailed or comprehensive systems investigation and design for the user. Neither the charges involved nor the bureau's staff establishment are likely to be such as to make this a viable proposition except in isolated cases. If a user wished to employ a service bureau without having his own systems analysts, he should commission a computer consultant to act in this capacity. Care must be taken when choosing the consulting firm that its staff has the practical experience of actually implementing work on a computer.

As stated earlier, bureau service is also used by some organizations as an introduction to data processing. This can be an economical way of learning the art, but the user must make sure that this is in fact happening and that his staff are not merely "spectators." It is not uncommon for a bureau user's staff to be genuinely convinced that they "know the ropes," whereas in reality they have acquired only a superficial knowledge of the techniques involved.

Accessibility to Files. In this context, accessibility means the lack of remoteness of the files. Files that are attached to a user's on-site computer are obviously the least remote and the most readily accessible. An on-line time-sharing system also provides good accessibility, but other facilities may not do so. This lack of file accessibility is often of no consequence for isolated applications, but may prove to be increasingly embarrassing in more comprehensive systems. The problem is related to both frequency of need to have access to the files and intervals between the user's contacts with the computer. In circumstances where file information is required at short notice, printed copies of the files have to be prepared at regular intervals and then held on the user's premises. Although this procedure may be satisfactory for low activity and nonvolatile master files, for other files there is a danger that the printed copies may be out of date when consulted.

12.4 SYSTEM PRESENTATION AND MANAGEMENT DECISIONS

After the design and appraisal of the data processing system, it should be formally presented to top management before implementation commences. Depending upon the decisions made at the feasibility study stage, the presentation might involve top management in arriving at a decision whether or not to proceed with data processing. In any case, even when a definite decision to proceed was made earlier, the sum of money and the reorganization inherent in this step make it vitally important that top management is fully aware of its consequences and benefits. It is also very probable that although the decision to go ahead with data processing has been made previously, the precise machine and facility have not been chosen.

It is unfortunate, but nevertheless true, that a computer is one of the most difficult of machines to really understand and yet one of the easiest to be impressed by. Although, in the final analysis, top management sanctions the order for a computer, the decision as to its identity must be made by the computer specialist. Only the company's data processing staff is qualified to arrive at the computer's configuration, and their opinion in the selection of the particular model should carry the utmost weight. In special circumstances, for instance, due to intertrading, top management may decide to place the order with a certain manufacturer. Nevertheless, the configuration of the computer should be specified by the user's systems analyst and not by the manufacturer.

Top management is truly concerned only with decisions about the output of the data processing system and its effect on the company; upper-level managers must not allow themselves to become deeply involved in arguments about hardware. The computer manufacturers are experts at impressing directors and high-level executives. Their electronic showrooms have been created to provide elegant demonstrations of the computer's capabilities, and this, combined with sumptuous entertaining, may be so overwhelming as to leave the prospective customer with a distorted view of the relative merits of the hardware.

What can the systems analyst do in order to give management a clear picture of the situation? As mentioned in Section 9.4, a systems definition ought to be prepared in the design stage. This definition is intended not only to instruct the staff who will implement the system, but also to clarify some sections that are suitable for assimilation by nontechnical staff such as top management. The systems analyst can also arrange discussions and talks as methods of data processing indoctrination.

Discussions. These are extremely useful to the systems analyst for putting forward ideas about the proposed system and for receiving criticism of it. Having listened to critical comments during the discussions prior to the formal presentation of the system, he can dispose of them either by adjustments to the system or by further explanation. Discussions help to engender a feeling of participation among top management, and also encourage the neophyte to try to understand the system.

Talks. Talks are given by the systems analyst for the benefit of all levels of management, and should be brief, informal, informative, and interesting. His main aim is to make management feel that the new system is a step toward the ideal means of control and that it can be developed logically in this direction.

Certain things are better explained verbally than in writing. In this category are the deliberate omissions from the system and from the recommended hardware—management does not like to feel that its associates or competitors have more advanced ideas. Talks provide an excellent opportunity to explain not only the proposed system and hardware, but also the advantages of these in comparison with alternatives. For instance, visual display units may have been installed by an associated company and have taken the company president's fancy; a rational explanation that these devices have been considered but found unsuitable puts his mind at rest.

Management Decisions

Top management, having been formally presented with details of the proposed system and still undecided, must finally arrive at a decisive conclusion. This may be (1) to go ahead as per the recommendations, (2) to make amendments to the system's output or objectives, (3) to postpone the implementation of the system for a stated period, or (4) to reject the proposed system. Conclusion (2) results in reconsideration at a later date of the amended recommendations. The reasons for conclusions (3) and (4) should, if possible, be made known, although some reasons—such as pending mergers, takeovers, or top-level policy changes—may make this difficult.

The situation to be avoided is disinterest or permanent indecision on the part of top management. This is demoralizing for the systems analyst, and also leaves a feeling of apprehension in the minds of other staff.

12.5 EXERCISE

Problem 1. Cost Comparisons

Draw up a cost comparison table for the years 1975 through 1979 to cover the following situation:

The PQR Company intends to rent a computer from the beginning of 1975 at a monthly rent of $5000 in order to carry out stock control, sales accounting and production control. The annual cost of punched cards is estimated at $3000, and the rent of other machines such as card punches will cost $17,000 per annum. Other continuing costs, including additional reels of magnetic tape, are likely to amount to $8000 in 1975 and $2000 per annum thereafter. The cost of air conditioning equipment is $22,000; furniture, $4000; and the preparation of the computer building $35,000. These costs will be incurred in the 1975 budget.

The first application to be computerized is stock control, and by the start of 1976 there will be a saving of 12 stock clerks plus $16,000 yearly for the rent of various machines associated with stock control. By 1977 it is expected that stock-holding costs will be reduced by $10,000 yearly, and double this amount in 1978 and 1979.

Sales accounting is to be computerized by the start of 1977, thus saving 20 sales clerks. Additionally, the earlier collection of debts will result in a net saving of $9000 yearly by 1978.

The implementation of production control will save 15 production planners by 1978 and $150,000 in production costs.

The average salaries of sales clerks is $8000 yearly; of production planners, $10,000 yearly; and stock clerks, $6000 yearly.

The data processing staff salaries are expected to amount to $120,000 yearly from 1975 onward; on top of this there will be training costs of $10,000 in 1975 and $5000 yearly thereafter.

Miscellaneous other costs are estimated at $25,000 in 1975, $10,000 in 1976, and $7000 in 1977 onward.

Solution to Problem 1. It must be realized that this problem represents a vast simplification of the real situation. Many factors have been omitted or simplified, and no account is taken of the net present values of cash flows (Section 9.5). Nevertheless, the factors cited are included in Fig. 12.2 for the period 1975–1979, and the results represent a NPV of $135,000 at a discount factor of 10 percent, or

$$-304,000 - \frac{129,000}{1.1} + \frac{44,000}{1.1^2} + \frac{363,000}{1.1^3} + \frac{363,000}{1.1^4} = 135,000$$

DATA PROCESSING - ADDITIONAL COSTS, $

	1975	1976	1977	1978	1979
Computer rental	60000	60000	60000	60000	60000
Punched cards	3000	3000	3000	3000	3000
Anc. M/Cs rental	17000	17000	17000	17000	17000
D.P. salaries	120000	120000	120000	120000	120000
Magnetic tape	8000	2000	2000	2000	2000
Air conditioning	22000	—	—	—	—
Furniture	4000	—	—	—	—
Site Preparation	35000	—	—	—	—
D.P. training	10000	5000	5000	5000	5000
Misc. costs	25000	10000	7000	7000	7000
Total	304000	217000	214000	214000	214000

DATA PROCESSING - POTENTIAL SAVINGS, $

	1975	1976	1977	1978	1979
Stock holdings	—	—	10000	20000	20000
Stock clerks	—	72000	72000	72000	72000
Stock ctl. M/Cs.	—	16000	16000	16000	16000
Sales clerks	—	—	160000	160000	160000
Debt savings	—	—	—	9000	9000
Prod. planners	—	—	—	150000	150000
Prod. costs	—	—	—	150000	150000
Total	—	88000	258000	577000	577000
Net savings	-304000	-129000	44000	363000	363000
Acc. net savings	-304000	-43000	-389000	-26000	337000

Fig. 12.2   Solution to Problem 1

249

CHAPTER 13

IMPLEMENTATION OF DATA PROCESSING SYSTEMS

13.1 SYSTEM TESTING

Before bringing the data processing system into use, it is of vital importance that it be both comprehensive within its intended limits and fully correct. Each program will by now have been written according to its specification (Section 10.6), and tested by the programmer to his complete satisfaction. The final responsibility for the correctness of both programs and system as a whole lies with the systems analyst. He must make absolutely certain that each run produces exactly what is required of it, and that the runs link smoothly together to provide the correct output of each routine. The linking together of runs during the testing procedures insures that the output of each processing run is fully correct as far as it relates to the input of the run that follows it. The linking together of processing runs during the systems-testing phase is often bedeviled by the fact that different programmers have written the programs within the one routine without having sufficient contact among themselves to insure an absolutely perfect interface.

Ideally, one set of basic data can be used to test all programs in a routine. In practice, this method is not usually suitable because test data chosen to extend one program's capabilities does not necessarily do so for another; there is often a conflict between the requirements of the respective data. The programmer will have created his own test data; the systems analyst should create another quite separate series based on actual source data so that a double check is made.

Characteristics of Test Data

There are six requisites to be satisfied:

1. Data must extend the run to its limits with regard to factors such as:
   (a) Sizes of input and output fields.
   (b) Sizes of calculated intermediate values.
   (c) Variations in the possible formats of indicative fields.
   (d) Detection of errors by feasibility checks, and the consequent action by the computer.
   (e) Variations of program paths, although it may not be possible to test all combinations of these at one time.
   (f) Interpretation of special symbols and coded fields.

2. The capacity of storage areas allocated to tables, indexes, files, and results must be tested against the maximum amounts and volumes of these data. This is especially applicable to storage areas in direct access files from which the overflow of records is possible. Although no system can be expected to handle excessive overflow, it is vitally important that errors caused by overflow be detected during the testing phase.

3. The testing of the internal organization of data is generally more difficult than the testing of calculations. Two quite separate lots of test data are therefore called for, each lot being designed to "stretch" the program or routine with respect to the characteristics being tested. In this context, data organization includes address generation or calculation, index creation and searching, overflow organization, and the creation of stored tabulations and lists for subsequent output.

4. Not only the visible output but also the data left stored on magnetic disks at the end of the run need checking. Although this data usually forms the input of another run, checking it at this point is not really a suitable way of verifying it. A special printout of this stored data, followed by a manual check, is necessary in order to insure that it is both accurate and complete. Nor is this stored data suitable for fully testing the run that uses it as input; a separate lot of specially created data is required for this purpose.

5. It does not follow that actual data is necessarily the most suitable for test purposes. The reasons for this are as follows:
   (a) Actual data is almost sure to be biased toward certain characteristics; for instance, a predominance of certain code numbers and the absence of others in a batch.
   (b) The output results from previous usage of the actual data were often not intended to be exactly the same as those expected from the new system. The reconciliation of the consequent different results introduces more work than the use of fresh data.
   (c) Actual data does not necessarily extend the program in the ways mentioned in items 1, 2, and 3 of this list.

6. Test data is also used to check that run times approximate those estimated. Actual data is convenient for this purpose because it is representative of what will be actually processed by the run.

Generating Test Data

One of the problems intrinsic to systems testing is that of providing enough data to simulate master files for program-testing purposes. If it is possible to create the actual master files before testing the other parts of the system, then the problem disappears. To do this is not always convenient, however, and in this situation dummy master files have to be created. Instead of punching and reading a large amount of artificial data, it is sometimes quicker and cheaper to create it by means of a "data generating" program. This is written in a generalized form and set up by parameters to suit the particular type of record that is required. There are limits to the degree to which this program should be made comprehensive, since it may become very complicated to write; nevertheless, for creating fixed- or variable-length record files, it can prove to be extremely useful. The factors involved in data generating are:

1. Each field in a record is of a given format and lies within maximum and minimum limits.
2. Each record consists of a given set of fields, some of which are fixed in number and others are variable within specified limits.
3. The records may be stored either sequentially according to a given key, or randomly, and are assigned to their respective locations in a direct access file by the normal file creation program (Section 13.2).

The data-generating program creates the fields of a record, one at a time, by generating pseudo-random numbers to form the fields. After being generated, the random number is checked for format and size; if acceptable, it is then inserted into the record as a field. This is repeated for each field of the record according to its parameters; when complete, the record is written into the file. If the records are being created for a sequential file, the key is generated within tighter limits in that it must be greater than the previous key but not too much greater. A suitable maximum increase is twice the average file increment; the latter is equal to maximum key *minus* minimum key in the file, divided by the number of records in the file. Alternatively, the keys can be generated *ad lib,* and the records then sorted into sequence before being written into the file.

13.2 CREATION OF MASTER FILES

As suggested earlier, the master files are the framework of a data processing system. It is therefore essential that they be created initially in a complete and accurate form. If this is not done, it is possible that some of the errors could remain undetected for a considerable time, with consequent errors in the output of the processing. In the more straightforward situations, the master file is created by reading cards or paper tape and simply copying the data onto tape or disks. In other cases the routines for file creation are complicated, even to the extent of being some of the most difficult jobs to program and operate.

Sources of Master File Data

It is quite usual for a number of different source documents to contribute data to the master file. These documents may well have been used in different departments of the company for many years, but may have been never previously brought together. This means that a lack of uniformity between documents almost certainly exists, including differences in items such as descriptions and code numbers. An example of this state of affairs could be in the creation of a master file for sales/stock processing, the stock levels being derived from the stock sheets, the sales history from the sales record cards, and the selling price from the sales catalog. Because of the diverse nature of the previous applications of these records, their nonuniform code numbers have been of no consequence. Contact between data from the records was formerly via human processing so that discrepancies were dealt with by memory and intuition.

It is commonly the case that items are present in one set of documents and missing from another set. This situation raises difficulties in the later stages of file creation if the file data is to be drawn from both sets; it is therefore essential that matching and correction runs be incorporated in the early stages of file creation routines.

File Creation from Subfiles. When a number of sources provide data for a master file, it is occasionally possible to combine the documents manually and then punch a length of paper tape or a batch of cards from each set of related documents. This is, however, a tedious process, both as regards manually collating the documents and punching from them. It is safer to keep the documents separate and to punch a subfile of cards or tape from each separate lot of similar documents. Each subfile is transferred onto magnetic tape or disks; at the same time, a proof list is prepared and control totals accumulated and printed. The subfiles are then sorted and merged to form the master file; any records for which data is missing are listed and omitted from the master file. Using this list of incomplete records, the missing data is found and inserted in an additional run.

An alternative method is to create the subfiles as above, but then merely to match the records without actually merging into one file. The result of this run is a printed list of omissions which is used to find the missing documents.

The subfiles are then corrected separately by insertion of the omissions, and finally merged to form one master file.

Proof Lists. The manual checking of a full detailed list of items in a large file can prove to be an overwhelming task. Nevertheless, it is useful to have a proof list available for checking individual items as and when necessary. With descriptive proof lists, it is sometimes possible to accelerate the checking process by merely scanning the items and checking against the source documents only those that appear suspect.

Control Totals. These are easily obtained from a computer run, but are not always easy to get in a manual procedure. In spite of this difficulty, it is worthwhile to obtain control totals from the source documents along the lines described in Section 11.4. If, as is likely, they do not appear as part of the existing system, they should be formed by using an adding machine for batches of not more than 2000 documents at a time. Thereafter, the control totals are accumulated and checked at every stage of the file creation procedure.

Time Spread of Master File Creation

It is frequently the case that a master file has to be created from source documents that are being continually amended. This is not so much the situation with cataloged data as with stock, sales, and production figures, for which there is little chance of finding a static situation. If master files could be created instantaneously, there would be little problem, but since they usually take a considerable time to create, the data for one record tends to pertain to a different date from that for another.

The essence of this problem is connected with record entering, and it must be remembered that once this has been done for a master record, all further amendments to that record must also be entered so that it will be absolutely up to date when the master file comes into use. Great care is necessary to insure that no amendments or movements are overlooked or duplicated after the date on which the master record is entered. A somewhat expensive method of achieving this end is to make photocopies of all documents as of a certain date and time; punching from these copies can then proceed at leisure, with later amendments being punched as and when they occur.

A more practical method is to enter a batch at a time in one run after normal working hours. This means that no changes can occur during this time, and amendments to each batch are clearly applicable from this date onward. As an extra safeguard, each record can contain initially its date of creation so that, provided amendments and movements are also dated, there will be no doubt as to their applicability.

13.3 CHANGEOVER PROCEDURES

The changeover from the existing system to the data processing system begins after the computer has been installed, but preparations for the changeover will have been going on for some time before this event. System testing should commence on other computers well before the installation of the user's own machine so that useful results can be produced immediately after its installation.

There are four basic alternative procedures for achieving the changeover; the one that is adopted in a particular situation depends upon the type of organization and the relationship of the old and new systems. Different procedures may be employed for the various applications within the one company. On the other hand, it does not follow that the same procedure should necessarily be employed for the same application in different organizations. The most suitable procedure is chosen by the systems analyst according to the prevailing circumstances.

Direct Changeover

This involves ceasing operations under the old system and immediately commencing the processing of data according to the new system. This is a somewhat drastic method and should be adopted only if no other procedure is suitable; for example, when it is impractical to meet the demands of extra work entailed in the other procedures. Whenever possible, direct changeover should be fitted into a weekend or statutory holiday period when work is slackest. This causes the minimum of disruption and allows the most time to bring the new system into operation.

One of the main disadvantages of direct changeover is the problem of insuring that the new system is functioning 100 percent correctly, since there is no basis of comparison with the previous system in that it did not produce the same results. Also, if results are found to be incorrect, it may be very difficult to retrieve the situation, especially if the old system cannot be re-employed temporarily until things are put right with the new system.

In the case of terminal-based real-time systems (for example, real-time banking), direct changeover may be the only feasible method. Certain situations allow for a combination of direct and stage-by-stage changeover (see below), each stage being in itself a direct changeover.

Direct changeover needs the most careful planning by the systems analyst, and the most diligent attention to timing by the operational staff. Given these requirements, the task can be accomplished without too much difficulty.

Parallel Running

With this procedure the current basic data is processed by both the old and new systems and, as far as it is practicable, the two sets of results are checked against each other. This is generally done by comparing all totals individually and making sample comparisons of the detail if the amount of output is extensive. Care must be taken to insure that not only the printed output of the new system is correct but also the carried-forward data.

Parallel running is normally carried out for one or two processing cycles. After the first cycle, the output from the old system is distributed; thereafter, the new output is distributed and the old output held in reserve in case of dispute. In cases where the new output has a very different format from the old, both sets of internal documents may be distributed together on the first occasion, accompanied if necessary by an explanatory note pointing out the changes. This arrangement is not advisable with externally distributed documents, but a brief explanation may be worth sending to the external recipients along with the new documents.

The major problem with parallel running is the duplication of work involved. In all probability the staff that has been working on the old system will also be involved in the new system; for this reason, parallel running must not be prolonged more than is absolutely necessary.

Pilot Runs

In this procedure, current data continues to be processed by the old system while previous data is reprocessed by the new system. Thus, provided the two systems are intended to produce similar results, a basis of comparison exists. The amount of previous data reprocessed by the new system depends upon the particular application; it is not usually necessary to reprocess all of a cycle's previous data, but care must be taken to insure that the chosen sample is truly representative. Provided the sample is not too extensive, there is no great hardship in pilot-running for several processing cycles. If an application consists of sections of data that have different characteristics, each such section may be used as pilot data in a different cycle. By this means, each section gets a check that is directed toward the particular characteristics of its data.

When the new system is proved to be correct, a double cycle in one period makes the pilot run into a parallel run. Thereafter, the old system can be abandoned in the knowledge that extensive checking of the new system has been carried out.

Phased Changeover

This is similar to parallel running except that initially only a portion of the current data is run in parallel on the new system; for instance, that pertaining to one department or section. During the following weeks more sections are transferred onto the new system, and in each case the old system runs in parallel for one processing cycle only. Thus, the old system is phased out as the new system builds up, and at each stage it is quite practical to check the new output against the old before distributing it. The total amount of extra work is generally less than that involved in parallel running.

Stage-by-Stage Changeover

In this context, a stage can be regarded as a routine or a complete application, and the method is not so much alternative as complementary to the methods described above. Stage-by-stage changeover is the arrangement adopted by nearly all companies for their major applications, often starting with the payroll, then stock recording, and so on. Stage-by-stage changeover can also be applied to the routines within an application. The most important factor to bear in mind with this method is the need to be able to dovetail the later stages without disruption of those already functioning.

A typical stage-by-stage changeover of a basic accounting system might be:

Stage 1. Preparation of invoices from dispatch.
Stage 2. Preparation of sales analyses from invoiced items.
Stage 3. Updating of sales ledger accounts from invoice totals.
Stage 4. Preparation of statements from the sales ledger accounts.
Stage 5. Automatic preparation of overdue account reminders, and so on.

13.4 INVOLVEMENT OF USER-DEPARTMENT STAFF

The data processing department cannot function in a vacuum; because it is essentially a service department, it must be employed by, and be in contact with, the other departments in the organization. The greatest benefit is derived from the data processing system when other staff are not only aware of its existence but have a genuine desire to make use of its services. This atmosphere is best created by the systems analyst in actively "advertising" the data processing system during its appraisal and implementation stages. This is less necessary if project teams, including user department staff, were formed to facilitate systems investigation, and if contact is maintained thereafter between the systems and the user department staff.

The staff from outside the data processing department can be regarded as falling into one of three groups:

1. Those who will be feeding data into the system and/or receiving results from it.
2. Management receiving results from it.
3. Other staff, whose work will be indirectly affected by the system.

These people, together with the organization's outside contacts, form the environment of the data processing system. Once the system is in operation, it may be too late to modify the environment if it proves to be confused or uncooperative. The systems analyst is well advised to encourage the right attitude among other staff prior to bringing the system into operation. This includes, for instance, agreement with the departmental managers as to who has ultimate responsibility for the correctness of documents going to outside organizations. In general, it is better for this responsibility to lie with the user departments, as they are geared to dealing with outside inquiries. This arrangement would not, however, absolve the data processing department from responsibility for the accuracy and promptness of its output.

Instructions to User Departments

It is the systems analyst's responsibility to insure that the user departments are instructed in the tasks that they have to perform in connection with the data processing system. He should also insure that they fully understand the purpose of the system's output that they receive. These points are covered by written instructions and explanations sent by the analyst to the heads of the user departments. The points will have been explained beforehand verbally; the written instructions are, in fact, a confirmation of what has already been agreed. The instructions and explanations are better expressed in advisory rather than authoritarian terms, and cover the following points:

1. The day and time by which each set of source data is to be ready.
2. Who is responsible for preparing each set of source documents, and to whom they are returned after being punched.
3. The preparation of input control totals and counts, exactly what these are, and where they are to be shown on the source documents.
4. The precise meaning of the information on each output document.
5. Who is to receive the output documents, the number of copies, day and time of completion.
6. A brief description of file contents so that user department staffs are aware of what information can be made available, and will not then start up their own private system to provide it.

7. What "on request" information (if any) is available from the system, with an indication of the waiting time for it.

8. Who to contact within the data processing department in case of difficulties with the source data or output, or when changes are desired. Although this person is usually the data processing manager or the senior systems analyst, it is better to formalize the arrangement so that the user departments do not make unofficial arrangements with programmers or operators. There is, of course, no harm at a later stage if these people become involved in discussing minor changes, provided their conclusions are agreed upon and formally included as amendments to the system.

Top Management Involvement

Top management should, to some extent, be involved in the implementation of the data processing system; there are several reasons for this:

1. Their advice and authority may be needed, for instance, in order to exert pressure on managers whose departments are laggard and endanger the implementation schedule as a consequence.

2. So that they can comprehend the volume of work involved in planning and installing a data processing system, and will then give this consideration before requesting modifications to the output of the system.

3. They will acquire a feeling for data processing, and this encourages them to employ new developments in hardware and techniques in order to improve management control of the organization.

4. The reports presented to them will be understood more readily, and this will initiate the possible benefits to be gained from conversational mode systems. In the long term, this last point is of considerable importance because only the top managers themselves can decide what they would gain from being able to obtain *ad hoc* reports, etc., from a conversational mode system. The systems analyst must not prejudge this issue; the demand for it should be allowed to arise as a natural consequence of top management's familiarity with the capabilities of data processing systems. Conversational mode, interrogation systems, on-line visual display, and similar arrangements can be hazardous and expensive "toys" in the hands of management uneducated in their use.

System Presentation to Other Staff

This is intended to give all other staff (who are not directly concerned) a general picture of what is involved in data processing. Staff from the user departments may also attend these presentation talks if they do not feel cognizant with any aspect of the system. The talks are given by the systems analyst, possibly assisted by some of the user department staff, and are directed toward the removal of the apprehension that can easily become established in the uninformed mind. The points to be included are:

1. A brief description of the hardware and its method of use in order to dispel rumors and worries.
2. Advantages to be gained from the new system.
3. Emphasis on human participation and responsibility.
4. Answers to any relevant questions. Those relating to security of employment should be dealt with by the management rather than the systems analyst.

13.5 DATA PROCESSING STAFF RECRUITMENT AND TRAINING

Recruitment of Systems Analysts

It is necessary to recruit and train the data processing staff long before the computer is installed; this requirement applies especially to systems analysts and programmers. The systems analyst is involved in designing the system, both before and after the decision to proceed with data processing is made. In practice, it is unlikely that he will have completed more than a small proportion of the total work when the computer is ordered. Before installation of the computer, and during the succeeding months, the remaining systems design is done, and in all but the most elementary applications the analyst continues his work after the implementation of the basic system.

In this rapidly changing field there is wide scope for steadily improving the system by introducing new hardware and techniques as these become available. It is therefore advantageous to set up a systems team on a permanent basis; its size depends upon the complexity of the organization's control and information system, and varies from one person to several dozen. When building up a systems team, its structure and growth rate will be based on the advice given by the first systems analyst. He will become a key man in the company; consequently, his caliber is of the utmost importance. In many cases, this man is later promoted to be the data processing manager or management services manager.

In some circumstances, the analyst who makes the initial recommendations is employed on a consulting basis, either by an outside consultant firm or within the organization, but not as a permanent systems analyst. If a team is created, there is room in it for analysts of lesser experience. These people will have the opportunity of working alongside the more experienced ana- lysts, and can thus be trained and gain experience while working in the organization.

Training of Systems Analysts

A variety of courses are now available for the training of systems analysts, run mostly by the manufacturers, colleges, and consultant firms. Manu- facturers' courses tend to be available only for staff from their customers, and like the consultant firms' courses, are rather expensive.

Recruitment of Programmers

The caliber of the programmers required depends not so much upon the complexity of the work to be programmed as on the level at which the systems analyst defines it. This level is to some extent flexible, but the norm should be taken as that described in Chapter 10. It is the responsibility of the analyst to decide this level when he has had an opportunity to assess the programmers' capabilities.

The area for programmer recruitment is best made quite wide; although a business background is a useful attribute, it is by no means essential. This means that potential programmers can be drawn from all departments within a company, including those in the factory as well as the office. Similarly, they can come from a wide range of outside organizations, including other com- panies or straight from school, college, or university.

The essential qualities to be looked for in a potential programmer are:

1. *Education:* Up to junior college level in one or two subjects, preferably of the logic type such as physics, statistics, mathematics, chemistry, and possibly languages.

2. *Aptitude for programming:* In some ways this is more important than educational qualifications, particularly if the recruit does not wish to go be- yond programming in his career. There are a number of programming apti- tude tests available, mostly run by the manufacturers. The results of past tests have been shown to correlate quite well with the subsequent efficiencies of the candidates as programmers.

3. *Desire to become a programmer:* It is difficult for a newcomer to data processing to comprehend what programming is about. It is therefore advis- able to give applicants some idea of the nature of the work before they under- go the aptitude test and training. Because a person is successful in the aptitude test, it does not automatically follow that he or she will be content to spend several years in the occupation of computer programmer.

Whenever possible, a proportion of new programming staff, say a quarter, should, when recruited, have had experience of writing programs in the language(s) to be used. This experience is valuable, since it not only accelerates implementation of the work to be done on the computer, but also enables raw recruits to supplement their training with advice from the more experienced staff.

Training of Programmers

Many programmers are trained by the manufacturers in full-time courses lasting about four weeks. A course of this type is directed toward one particular programming language that can be used with the manufacturer's computers. These are the most suitable courses for programmer training because they are full-time courses carried out in a computing environment and run by the originators of the relevant language.

Colleges often run part-time courses in the higher-level languages such as COBOL; these courses are suitable for gaining a general appreciation of programming, at moderate expense.

In the case of courses run by other less well-known organizations, the prospective user is advised to check their credentials most carefully before entrusting his staff to their training methods.

Recruitment and Training of Other Data Processing Staff

The recruitment and training of computer operators does not normally pose a problem. They should have high school education and be capable of working to tight schedules and of following written instructions meticulously. Operators do not require a knowledge of programming, but it is probable that some of them will eventually become programmers after being retrained. Senior operators (shift leaders) must be capable of controlling staff and of accepting responsibility for the computer's throughput. An operator's training is usually carried out on site by the more senior operators, or initially by the manufacturer's staff.

The tape librarian and data control staff do not need to attend formal training courses. Their main attributes are the ability to organize and administer the control of files and of documents entering and leaving the data processing department.

13.6 IMPLEMENTATION PLANNING USING NETWORK ANALYSIS

Network analysis is also known as "critical path method" and by other names, and in its more sophisticated form as PERT (Program Evaluation and

Review Technique). It is widely employed in business and industry for project planning, and since its inception in 1957, a large number of variations of network analysis have been developed, mostly to meet the requirements of particular industries or specialized applications. In its original form, the emphasis was on the control of projects from the time aspect, but other aspects have now become equally prominent, such as resource planning, cost control, and multiproject scheduling. Readers requiring information on these aspects of network planning are advised to consult the wide range of literature available on PERT.*

When employed as a technique for planning the implementation of a data processing system (the project), network analysis is mainly concerned with the time control aspect. That is, it covers the interrelationship of all the necessary jobs (activities), their estimated time durations, and the progressing of their achievement. Although implementation planning is a relatively small type of project, and like most other projects it can be accomplished without the use of network planning, it is worthwhile employing the technique for a number of reasons:

1. This is an opportunity for the systems analyst and other staff to familiarize themselves with it. Any analyst should be capable of utilizing it, not only for implementation planning but also for other projects with which he will come into contact during his career.

2. Because the network can be analyzed by using the computer manufacturer's package, this provides early experience for the analyst in the use of packages and also of a service bureau, since this is likely to be used prior to delivery of the user's own computer.

3. Preparation of the network diagram and the estimation of activity durations force the analyst to consider carefully all the jobs involved in system implementation, with the result that there is less chance of any activity being overlooked or underestimated.

4. Calculations based on the network diagram indicate the activities that need the most attention if the scheduled dates, including the completion date, are to be maintained. Other less critical activities also have dates assigned to them so that there is no doubt as to when each job must be carried out.

Preparation of a Network Diagram

A typical but simplified network diagram as applied to the implementation of a data processing system is shown in Fig. 13.1. This diagram does not purport, however, to cover all situations; each project must be planned on its own merits so as to include its own particular activities. Each activity is

---

*Woodgate, *Planning by Network,* Auerbach, Philadelphia, 1971.

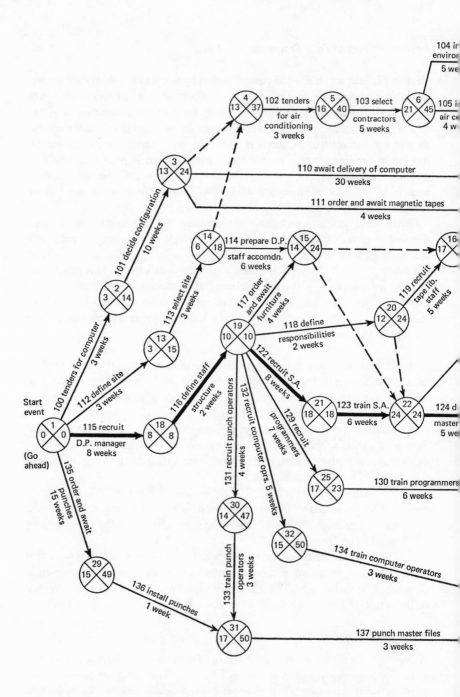

Fig. 13.1 Network diagram for implementation of a data processing system

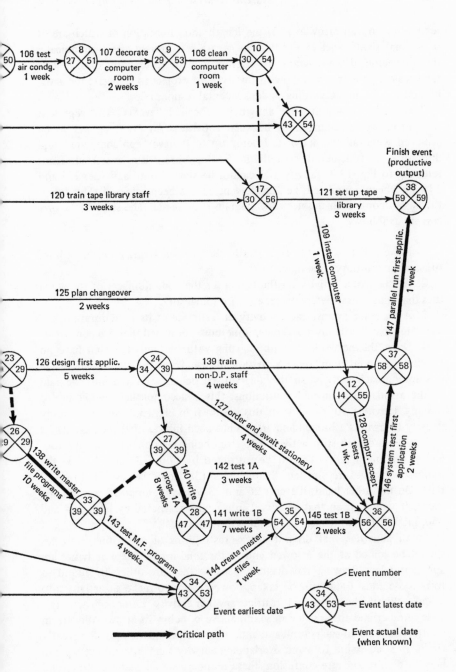

Fig. 13-1 (Cont'd)

265

represented by an arrowed line, the length and orientation of which are of of no real significance except that the general left-to-right direction indicates the time sense of the diagram. No attempt is made to draw the diagram to a time scale, nor is it advisable with this type of project to segregate the activities on the diagram according to departmental responsibility.

The circles separating the activities are called "events" and represent the chronological connections between activities. An event is said to be achieved when all the activities leading up to it have been completed, and until an event is achieved, no activities leading from it can be started. Thus, referring to Fig. 13.2, activity 137 cannot be started until activities 133 and 136 have been completed; i.e., until event 31 has been achieved. This rule is the basis upon which the network diagram is prepared. The steps in the process are as follows:

1. Make a list of all the activities in the project, giving each one a brief title and an arbitrary code number.

2. Against each activity on the list, enter the code numbers of the activities that must immediately precede it chronologically.

3. Also, enter against each activity an estimate of its time duration; the estimates may be in any convenient time units, provided they are consistent.

4. Draw the network diagram, starting with the activities that have no restrictions upon them; these activities all lead from the start event. When positioning each activity on the diagram, particular attention must be paid to the avoidance of invalid restrictions; this is accomplished by the use of dummy activities. An example of this is shown in Fig. 13.2, where a dummy activity (drawn as a broken-line arrow) has been inserted between events 23 and 26 in order to stop activity 126 from being restricted by activity 130, and yet allow activity 138 to be restricted by both 124 and 130. If the dummy activity were omitted, then activity 138 would be restricted by 130 only. On the other hand, if events 26 and 23 were superimposed to form one event, it would mean that activity 126 was also restricted by both 124 and 130, the latter of which is not a genuine restriction.

5. All activities that do not act as restrictions on any other activities should be joined at the arrowed end to the final event. These activities are apparent when drawing the diagram because they are otherwise found to form loose ends. Care must be taken, however, not to make an activity into a final activity when actually it restricts another activity. Enter each activity's code number and title on the diagram, above or below its arrow. Allocate an arbitrary number to each event, and enter it in the top quadrant of the event's circle. It is advisable to avoid overlapping activity numbers and event numbers so as not to cause confusion; these numbers are all quite arbitrary and have no connection whatsoever with each other.

	Activity	Est. Time	Dates				Floats			
Code	Description	Weeks	ES	EF	LS	LF	TF	FFE	FFL	IF
* 115	Recruit D.P. manager	8	0	8	0	8	0	0	0	0
* 116	Define D.P. staff structure	2	8	10	8	10	0	0	0	0
* 122	Recruit systems analysts	8	10	18	10	8	0	0	0	0
* 123	Train systems analysts	6	18	24	18	24	0	0	0	0
* 124	Design master files	5	24	29	24	29	0	0	0	0
* 138	Write master file programs	10	29	39	29	39	0	0	0	0
* 140	Write batch 1A programs	8	39	47	39	47	0	0	0	0
* 141	Write batch 1B programs	7	47	54	47	54	0	0	0	0
+* 145	Test batch 1B programs	2	54	56	54	56	0	0	0	0
* 146	System test first application	2	56	58	56	58	0	0	0	0
* 147	Parallel run first application	1	58	59	58	59	0	0	0	0
+ 142	Test batch 1A programs	3	47	50	51	54	4	4	4	4
126	Design first application	5	29	34	34	39	5	0	5	0
129	Recruit programmers	7	10	17	16	23	6	0	6	0
130	Train programmers	6	17	23	23	29	6	6	0	0
117	Order & await furniture	4	10	14	20	24	10	0	10	0
+ 143	Test master file programs	4	39	43	49	53	10	0	10	0
+ 144	Create master files	1	43	44	53	54	10	10	0	0
100	Accept computer tenders	3	0	3	11	14	11	0	11	0
101	Decide configuration	10	3	13	14	24	11	0	0	0
110	Await delivery of computer	30	13	43	24	54	11	0	0	0
109	Install computer	1	43	44	54	55	11	0	0	0
128	Computer acceptance tests	1	44	45	55	56	11	11	0	0
112	Define computer site	3	0	3	12	15	12	0	12	0
113	Select computer site	3	3	6	15	18	12	0	0	0
114	Prepare D.P. staff accomdn.	6	6	12	18	24	12	2	0	0
118	Define staff responsibilities	2	10	12	22	24	12	0	12	0
127	Order & await stationery	4	34	38	52	56	18	18	13	13
139	Train non-D.P. staff	4	34	38	54	58	20	20	15	15
102	Accept air condg. tenders	3	13	16	37	40	24	0	0	0
103	Select air condg. contrs.	5	16	21	40	45	24	0	0	0
104	Install environment	5	21	26	45	50	24	0	0	0
106	Test air condg. plant	1	26	27	50	51	24	0	0	0
107	Decorate computer room	2	27	29	51	53	24	0	0	0
108	Clean computer room	1	29	30	53	54	24	0	0	0
105	Install air condg. plant	4	21	25	46	50	25	1	1	0
121	Set up tape library	3	30	33	56	59	26	26	0	0
125	Plan changeover	2	24	26	54	56	30	30	30	30
131	Recruit punch operators	4	10	14	43	47	33	0	33	0
133	Train punch operators	3	14	17	47	50	33	0	0	0
137	Punch master files	3	17	20	50	53	33	23	0	0
135	Order & await punches	15	0	15	34	49	34	0	34	0
136	Install punches	1	15	16	49	50	34	1	0	0
132	Recruit computer operators	5	10	15	45	50	35	0	35	0
134	Train computer operators	3	15	18	50	53	35	25	0	0
119	Recruit tape library staff	5	12	17	48	53	36	0	24	0
120	Train tape library staff	3	17	20	53	56	36	10	0	0
111	Order & await mag. tapes	4	13	17	52	56	39	13	28	2

NOTES

* Critical path (shown as heavy lines).

+ On bureau computer prior to installation of user's computer.

Fig. 13-2 Activity dates and floats for network in Fig. 3.1.

7. Enter the estimated time of each activity below its arrow. Dummy activities are not numbered and normally have zero time.

The network diagram is now ready for use. The calculations inherent in using the diagram involve quite simple arithmetic, but owing to their large volume and the intricate relationship of activities, it is easy to make a mistake.

If there are more than about a hundred activities in a project, the reader is advised to make use of a computer PERT package so as to avoid the burdensome arithmetic involved. With a project for the implementation of a data processing system, the number of activities is not likely to be large, but nevertheless it is worthwhile to employ a package in order to gain experience in their use. The calculations done by the package can be paralleled manually, thus enabling the user to acquire a detailed understanding of network analysis.

Critical Path

It is obvious from even a superficial inspection of a network diagram that a large number of different paths go from its start to its finish. The project as a whole is incomplete until every activity on every path has been carried out. Each path's activities are in series, which means that the minimum estimated time for a given path is the sum of the estimated times of its activities. Owing to the cross connection of paths, it is highly improbable that a path can be completed in its minimum time, since its activities are delayed by other connecting activities. This implies that because one or more activities on a path are delayed, other of its activities have spare time. This argument applies to all paths except the one for which the sum of its activities' times is the largest in the network. This is called the "critical path" and is the one that determines the overall project estimated time.

The activities on the critical path are termed "critical activities," and these are the ones that must be accomplished within their estimated times if the project is to be finished by the estimated date. In Fig. 13.2, the critical path is shown as a heavy line, and as seen, it is possible for it to include several dummy activities.

Identification of the Critical Path. Bearing in mind that an event is achieved when all activities leading up to it have been completed, it follows that the earliest achievement date for an event is that which allows for completion of the longest time path up to it. Thus, in the example in Fig. 13.1, consider event 15; the two paths leading up to it are activities 112, 113, and 114, taking 12 weeks; and activities 115, 116, and 117, taking 14 weeks. Since both paths must be completed, the earliest date for achieving event 15 is week 14.

Each event's earliest date is entered in the left-hand quadrant of its circle. The method for finding all the earliest dates is to enter zero or some other starting date as the event's earliest date, and then to proceed forward through the network, entering earliest dates whenever it is possible to do so. Each earliest date is equal to the preceding event's earliest date plus the intervening activity's estimated time. Where there are several events in process, an earliest date is calculated for each and the latest of these is used.

Thus, referring to event 34, there are three activities leading from three preceding events and so the set of calculations for event 34 are:

1. Earliest date of event 33 (= 39) plus estimated time of activity 143 (= 4), giving 43.
2. Earliest date of event 31 (= 17) plus estimated time of activity 137 (= 3), giving 20.
3. Earliest date of event 32 (= 15) plus estimated time of activity 134 (= 3), giving 18.

The latest of these is given by 1, so the earliest date for event 34 is week 43.

Having calculated and entered all the earliest dates, the project's estimated time is now known; i.e., the difference between the finish event's earliest date and the start event's earliest date. Because the determination of the finish event's earliest date has taken into consideration all paths through the network, including the critical path (although this is not yet identified), this date cannot be exceeded (as an estimation). The earliest date of the finish event is therefore the same as its latest date.

The procedure is now to work backward through the network, entering the latest dates of the events (in the right-hand quadrants). The latest date of an event is the date up to which its achievement can be delayed and yet still leave time for all its following activities to be accomplished without delaying the project's completion date. The latest date of an event is equal to the latest date of its succeeding event, minus the intervening activity's estimated time. Where there are alternatives, the earliest of the calculated latest dates is selected. For example, the calculations pertaining to event 3 are:

1. Latest date of event 4 (= 37) minus time for dummy activity (= 0), giving 37.
2. Latest date of event 11 (= 54) minus estimated time of activity 110 (= 30), giving 24.
3. Latest date of event 17 (= 56) minus estimated time of activity 111 (= 4), giving 52.

The earliest of these is given by 2, so the latest date for event 3 is week 24.

This procedure continues until the start event is reached, the latest date of which should also be calculated so as to prove that it is the same as its earliest date. This is always so, and by arriving at this conclusion logically, it is probable that no mistakes have been made in the other events' dates.

It will now be observed that some events have earliest and latest dates that are the same as each other; these events are said to have zero "slack," and they all lie on the critical path. An event's slack is equal to its latest date minus its earliest date and the critical activities are those that lie on the path through all the zero-slack events (the critical path), as can be seen in Fig. 13.7. Care should be taken not to erroneously include as critical any activities that join two zero-slack events but which bypass others; for example, activity 125.

It is theoretically possible to have two or more critical paths in a project, but in reality this is highly unlikely. More probable are short lengths of parallel critical path, and although there are none in Fig. 13.2 as it stands, this would be the case if activity 125 had an estimated time of 32 weeks; i.e., equal to the sum of the times for activities 124, 138, 140, 141, and 145.

Activity Floats

As stated earlier, critical activities are those for which a delay in their completion causes a delay in the completion date of the overall project. Other, noncritical activities have associated with them a certain amount of spare time, known as "float." Within limits, the times actually taken to carry out noncritical activities can exceed their estimates without delaying the overall project; this is because the excess times are absorbed by the floats.

Each activity's floats are determined by the dates of its preceding and succeeding events. Since there are four combinations of these dates, there are four different floats applicable to each activity. If for any activity,

$$A = \text{its estimated time}$$
$$PE = \text{earliest date of its preceding event}$$
$$PL = \text{latest date of its preceding event}$$
$$SE = \text{earliest date of its succeeding event}$$
$$SL = \text{latest date of its succeeding event}$$

then its four floats are

$$\text{Total float} = SL - PE - A$$
$$\text{Free float early (free float)} = SE - PE - A$$
$$\text{Free float late} = SL - PL - A$$
$$\text{Independent float} = SE - PL - A$$

When the calculated value of a float is negative, it is regarded as zero. This can be seen in Fig. 13.3, which gives the floats pertaining to the network in

Fig. 13.2. The four floats are meaningful only in relation to the activity's contiguous activities:

1. *Total float* is a measure of the absolute maximum spare time that can be allowed for an activity if its neighboring activities' dates are arranged to provide for this.

2. *Free float early* is the spare time that an activity can have if it is started as early as possible; if the activity is not started as early as possible, all or some of this float can be allocated to its preceding activities.

3. *Free float late* is the spare time that an activity can have if it is finished as late as possible; if it is not finished as late as possible, all or some of this float can be allocated to its succeeding activities.

4. *Independent float* is the spare time that is available only to the particular activity; it cannot be shared with adjacent activities, and represents spare time that cannot be turned to advantage in the planning of activity dates.

Activity Dates

Each activity is surrounded by two events, each of which has an earliest date and a latest date (although these may sometimes coincide). Thus, an activity has a limit imposed on the date at which it is able to start; i.e., its earliest start date is the earliest achievement date of its preceding event. There is also a limit on an activity's latest finish date; i.e., the latest achievement date of its succeeding event. The noncritical activities, which have float, can be planned to take place at any time within their limits, but it must be remembered that activities' dates are mutually dependent except for those pertaining to activities with independent float.

Using the same notation as above, the four dates applicable to each activity are

$$\text{Earliest start date} = PE$$
$$\text{Earliest finish date} = PE + A$$
$$\text{Latest finish date} = SL$$
$$\text{Latest start date} = SL - A$$

Care should be taken with dates and week numbers to insure that a consistent relationship is used. In the example, date 0 means the beginning of week 1; date 7, the beginning of week 8 (or end of week 7); and so on. This relationship is convenient because we can start at date zero, but in actual practice it is likely that calendar dates are used because these are more akin to real-life usage.

Taking activity 120 as an example, this could start at the beginning of week 18 because this corresponds to its preceding event's earliest date (17), and could therefore be carried out in weeks 18, 19, and 20. On the other hand, it could possibly be delayed until weeks 54, 55, and 56 (corresponding to the latest date of its succeeding event; i.e., date 56). There are, of course, numerous other weeks during which this activity could be implemented; and assuming it cannot be split, there are in fact 37 sets of 3 weeks.

Progressing of Projects

Network planning is not intended to apply merely to static situations. The initial network having been drawn, and the activity dates and floats calculated, these are then used as a means of keeping a check on the progress of the project throughout its life. When activities are completed—completely or partially—the network is re-analyzed so that critical and near-critical activities are detected and can be closely watched. It is, of course, quite possible that the critical path will change its course several times during the life of a project. This is brought about by the insertion and deletion of activities, by amendments to estimated times, and by the actual dates of completion of activities.

If a computer package is employed to analyze the network, the current position of the project is held on magnetic tape or disk, and is updated at regular intervals when amendments occur. It is not necessary to redraw the network diagram at any time unless very drastic changes are made to it. The outputs of packages are very flexible, the information being producible in various sequences so as to emphasize the main points of interest in a particular project. In addition to the conventional listed output such as that in Fig. 13.3, the activity dates and floats can be printed by the computer in the form of bar charts. The list in Fig. 13.2 is in earliest start within total float sequence; this means that the most critical activities appear at the beginning of the list and are thereby most easily recognized by the project controller.

When using network analysis for system implementation, a careful inspection of the progress reports enables the systems analyst to anticipate difficulties in relation to the completion dates of the activities therein. This means that measures can be adopted to avoid hold-ups, and thereby insure that the data processing system will commence operation on time.

In the preceding project it would be the critical activities (asterisked), shown at the top of the list in Fig. 13.2, that need most attention. Thereafter, the activities in the list are of decreasing significance from the project completion date aspect.

In practice, it would be necessary to update the project data regularly (say, weekly) throughout the project's life. Let us assume that this is done and that by date 36 the situation is as follows:

Activities 106, 111, 124, 125, and 132 are completed.

Activity 110 has 13 weeks to completion, activity 120 has 2 weeks, activity 134 has 1 week, activity 137 has 2 weeks, and activity 138 has 4 weeks.

Each completed or partially completed activity implies that all its preceding activities have been completed; this follows from the logic of network layouts, as explained earlier. For instance, the completion of activity 106 means that activities 100-105, 112, and 113 must also have been completed, since they all logically precede 106. If one is doing the network planning manually, the network diagram would now be redrawn as in Fig. 13.3. As can be seen, the completed activities have been omitted and the partially completed activities have been omitted and the partially completed activities have been drawn

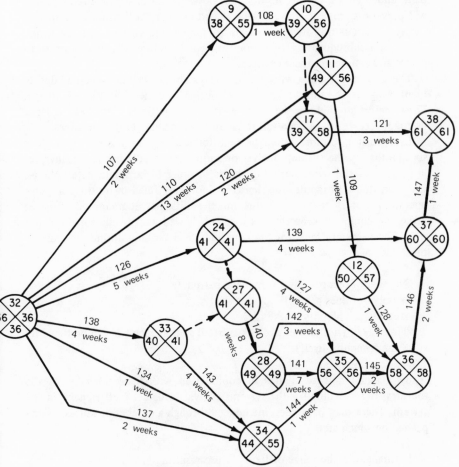

Fig. 13.3 Network diagram at date 36

joined together so as to create a new "start" event. The event dates are then recalculated and entered in the circles as previously, and the event numbers are the same as before, event number 32 being taken as the new "start." Note that the estimated finish date has been delayed by 2 weeks, and that a new critical path has appeared.

13.7 SYSTEM MONITORING

The purpose of system monitoring is to keep a watch on the efficiencies of the data processing routines by making regular checks after their implementation. Although a properly designed data processing system always produces complete and accurate output results, it is possible that changing circumstances will prevent it from doing so in an efficient manner. At the time of designing the system, it is not always possible to foresee the changes that can come about, and subsequent system monitoring is therefore advisable in order to detect and counteract these changes.

The two main factors with which the systems analyst is concerned during monitoring are the times taken by the routines and the utilization of storage. These factors are tied together closely in that an inefficient use of either serial or direct access storage may well show up as an increase in the run times.

Every major routine should be monitored at regular intervals, say, every six months, by the systems analyst or another assigned person. This involves inspection of the computer's log sheets in order to correlate data volumes with run times. Computer log sheets are normally filled in by the computer operators, but it does not require much additional programming effort to make the computer keep its own log. This is well worthwhile for the major runs. The necessary facts for the computer to determine during each run are those such as:

Number of records input from each peripheral.
Number of lines printed.
Activity or volatility of the files.
Time taken (if a real-time clock is fitted so that each run can be timed by the computer itself).

By examining these figures in conjunction with his knowledge of the routines, it is possible for the systems analyst to detect inefficiencies in the system. These may have been introduced through a number of causes, principal among which are:

1. Changes in the range of items in use resulting in many unused records

being left in the files, such as disused parts, obsolete products.
2. An overloading of a direct access file, giving rise to an excess of overflow records and longer run times.
3. The introduction of new items, with unsuitable code numbers, causing the address generation algorithms to become less efficient and thereby lowering the uniformity of record distribution in a random file.
4. Omissions of output results caused by deliberate or unintentional operating errors.

As well as the preceding technical aspects of system efficiency, there are a number of other factors entering into the maintenance of standards. These should, of course, be considered in relation to changes in the organization's modus operandi and to changes in government policy (for example, taxation). Nevertheless, once the system has become established as operational, the systems analyst should look for its promised benefits and also for unexpected contingencies. Among these are:

1. Unplanned increases in data processing staff and hardware.
2. New projects not being implemented to schedule.
3. Expected clerical staff reductions not achieved.
4. Unexpected difficulties in achieving acceptance of the new system by user departments.
5. Anticipated benefits not realized, such as stock-holding reduction and faster production throughout.
6. Additional work in systems design and programming brought about by government requirements and by company changes.

Contact with the Other Departments

Another aspect of system monitoring is the maintenance of contact between the systems analyst and the recipients of the system's output. It is essential that the objectives and output of the system are kept up to date, and not be allowed to lose their values through default. The degree to which contact is maintained depends upon the type of organization and the applications involved. In some cases there is inherently no change in an application for many years; in others the ability of the system to change course quite frequently is of great value. Although a data processing system should be designed to accept a fair degree of foreseeable changes, there may be circumstances that arise unpredictably and therefore call for special arrangements. These are discovered as a consequence of maintaining contact with the user departments, and asking the staff therein questions such as:

What further requirements or changes are needed in future?

Are any of the output results now redundant?

Are there any impending changes of consequence to the ranges of items?

As far as is known, will the input volumes increase significantly in the near future?

13.8 COMPUTER MANUFACTURERS' SERVICES

Following the signing of the contract for the purchase or rental of a computer, the user may have to rely heavily upon support by the manufacturer. This is especially true if the user has no previous experience of data processing. It is therefore important that the decision as to which manufacturer to order from is based not only on the characteristics and cost of hardware, but also upon the services that the manufacturer can offer. Having realized this, the user should find out from each prospective supplier what services are available, but at the same time endeavor to become self-supporting as soon as possible. A policy of leaning on the manufacturer and obtaining every iota of service—based on the philosophy of getting one's money's worth—is both risky and inefficient. The inevitable withdrawal sooner or later of this support invariably exposes a weak and inefficient data processing department.

What services can a user expect to receive, and to what extent are these covered by the contract? The latter point depends on what special arrangements, if any, are agreed upon between the manufacturer and the user. Generally, only computer maintenance is covered by the terms of the standard contract, and this is so because an additional charge is made for it, although maintenance charges may be also included in the rental. The other services described below are obtainable to different extents, depending on the manufacturer's ability to provide them and the user's ability to force the manufacturer to guarantee them.

Training Services

The computer manufacturers all have training schools and, generally speaking. these are the best available, particularly for lower-level courses. The prospective user should obtain the manufacturers' training course prospectuses, and from them discover the range, cost, and frequency of the courses held. The full spectrum of training includes courses required by top management, middle management, data processing management, systems analysts, programmers (elementary and advanced), and computer and key operators. By and large, these courses are held at the manufacturer's residential schools on a full-time basis, the period involved varying from one or two days for top management courses up to a few weeks for programming courses. Exceptions to this arrangement are courses that are run specially for one customer on his own premises—also operator training, which is usually also carried out locally. Self-tuition kits based on teaching machines and special literature are sometimes also available.

The overall cost of training staff at all levels can amount to a considerable sum, but set against this are the "free" or reduced-price courses that may be available to new users. The value of training discounts is normally based upon the value of the computer ordered.

Programming Support

This takes two forms. The first is actual program writing done for the user by the manufacturer's programmers. However, unless the value of the computer order is very large, it is unlikely that any free support of this type will be forthcoming. Nevertheless, there are chargeable programming services available from most manufacturers and from other programming service firms (software houses). The biggest problem in utilizing these services is communication, but provided the systems analyst specifies the runs as described in Section 10.6, or as per the service's specification requirements, then the service should be effective. These services tend to be expensive, but are useful for helping out with overloads of programming such as might be experienced in the early stages of implementation. A computer user would be unwise to employ a programming service to the complete exclusion of having his own programmers.

The second form of programming support is advisory; i.e., the manufacturer's programmers are available to advise the user's programmers during the first few months after the latter's initial training. The need for this support diminishes rapidly, but it can be extremely valuable if the user has no experienced programmers of his own. The prospective user should ascertain each manufacturer's ability to provide locally based advice from truly experienced programmers.

Specialized Advice

This is connected with the design of special systems and the employment of sophisticated techniques. These include the various operational research techniques, network planning and PERT, production control, and other application packages, as well as knowledge of the user's industry or organization. Although a manufacturer cannot be expected to have a comprehensive team of specialists in every local branch office, these people should be available from elsewhere at reasonably short notice and be prepared to make occasional visits to the user. If the user's systems analyst feels that specialized advice is vital to the system, its requirement should be written into the contractual agreement with the manufacturer. As in the case of programming advice, the specialist must be of the right caliber, with practical experience of implementing the application in which he specializes.

Program Testing

Prior to delivery of the user's computer, it is essential that sufficient programs be written and tested to enable the computer to produce useful results immediately after it is installed. To have it standing idle, apart from program testing, for perhaps several months is both costly for the company and damaging to the data processing department's prestige. Facilities must therefore be provided for program testing on another computer, preferably of a very similar configuration to the user's own computer. As with training, program testing quickly accumulates a substantial cost, and it is therefore wise to obtain a written statement from the manufacturer, specifying the facilities available and their cost. The factors involved in this matter are:

1. The configuration and location of the computer(s) available for program testing.
2. The arrangements for carrying out program testing; by whom, by what method, and, if done at a distant point by the manufacturer, the turnaround time between receipt of a program for testing and its return to the user.
3. The number of hours of free testing and the cost rate of subsequent chargeable testing.
4. The approximate time taken per individual test, and the relation between test runs and compilation. This and the preceding factor are of interest in an attempt to approximate the cost of preliminary program testing.

Installation Advice

The prospective user, once having acquired sufficient knowledge about computers to venture placing an order, finds himself thrust into the world of air conditioning and environment preparation. As with data processing itself and most other complex subjects, the first problem is to understand the jargon. Unfortunately, time does not allow for this, with the result that the user is soon confounded by the cross claims and varying specifications of the suppliers of air conditioning plant and environment equipment.

The computer manufacturer is ethically obliged to render assistance by making available his environment specialist to discuss the problems with the air conditioning suppliers' representatives. He should also advise on the form of tender and the best quotation for the user to accept.

Computer Maintenance

The computer, like other complex equipment, requires repair and regular maintenance; these apply particularly to its mechanical parts. The user should require the manufacturer to state how much time is required for scheduled maintenance, and arrange with him the times at which this will be carried out. Other significant factors relating to maintenance are:

1. The manufacturer's facilities for coping with machine faults and breakdown. If two or more manufacturers are involved in supplying connected equipment, the maintenance interface should, as far as is possible, be established from the start. This applies particularly to data transmission equipment, for which the responsibility for faults can so easily be tossed back and forth between the suppliers.

2. Where are the maintenance engineers based, and what is the ratio of computers to engineers in the area? If the user's computer is large, it is not unreasonable to expect to have a resident maintenance engineer.

3. What reserves of maintenance service can be called in if there is a major breakdown or persistent trouble?

4. What is the situation regarding spare parts; where are these held, and what is the maximum time taken to obtain any spare, including major components?

5. Is a downtime agreement included in the contract in order to recompense the user for losses caused by computer breakdown?

Standby Arrangements

The two main reasons for being interested in standby arrangements are (1) in case of persistent breakdown and (2) to cope with unforeseen temporary overloads. The latter tends to occur as a consequence of the former.

The most convenient standby arrangement is to arrive at an understanding with another local user who has a similar configuration. If his configuration is usable but not identical, it is worth preparing modified versions of the important programs to suit his configuration. Failing this arrangement, the manufacturer's computers are the next best bet. The questions to be answered here are:

What configurations are available and where are they situated?
Are these computers available solely for standby purposes?
What financial arrangements are involved?

Computers installed by the manufacturer for demonstrations and bureau service work are generally too heavily committed to be suitable for standby purposes.

Service Bureau Facilities

The reasons for employing a service bureau were expounded in Section 12.3; it is also common for users to employ bureaus for master file preparation (Section 13.2) during the system implementation phase.

The factors to be investigated in relation to proposed bureau service are:

1. Who owns the programs written by the bureau's programmers but paid for by the user? It does not always follow that the payment of perhaps several hundred dollars to have a program prepared entitles the user to have a copy of it for either use on his own computer or resale.

2. The quotations received from several bureaus, based on identical job specifications, turnaround times, etc., should be compared. These vary enormously, both with regard to programming charges, file creation and job setup charges, and regular running charges.

3. The quality of the service bureau should be discussed with other users of the bureau. This applies to turnaround time, accuracy, completeness and presentation of output, and general reliability. Also of interest is the continuity of employment of the bureau's staff; if this is short-lived, it is difficult for the user to build up his bureau routines because the bureau staff never becomes really familiar with his problems.

4. As with all contractual agreements, the prospective bureau user should read the small print on the back of the contract form most carefully. The conditions thereon may be most enlightening—if not to say alarming!

13.9 EXERCISES

Problem 1. Network Planning

In relation to the network diagram in Fig. 13.3, assume that by date 42, activities 108, 120, 123, and 137 have been completed, activity 126 has 1 week to completion, activity 110 has 10 weeks, activity 138 has 1 week, and activity 121 has 2 weeks. Redraw the network diagram and draw up a list as per Fig. 13.2 for the outstanding activities.

Solution to Problem 1. See Figs. 13.4 and 13.5.

Problem 2. Network Planning

It is required to implement an additional application within an existing

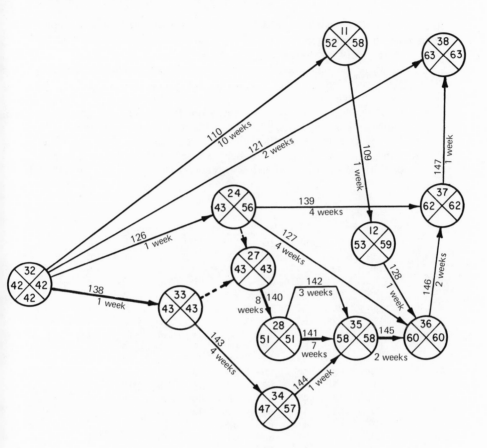

Fig. 13.4 Network diagram at date 42 (Problem 1)

data processing system. The jobs involved in so doing are a preliminary survey taking three weeks, followed by a system investigation taking six weeks, and then system design occupying eight weeks. Thereafter, there will be three sets of programs to be written and tested in parallel. Set A is estimated to take 12 weeks to write; set B, 8 weeks; and set C, 16 weeks. Program testing is expected to occupy a quarter of the respective writing times. After all programs have been tested, there will be a system test occupying three weeks. Following the systems investigation, there are four weeks for file design, followed by one week of user-staff training, and then five weeks for file creation. The latter must be finished before the testing of program set A can commence.

Immediately after the preliminary survey, it will be possible to place an order for additional equipment; the delivery period for this is 13 weeks plus 1 week for installation, which must be finished before testing of program set C can commence.

(a) Draw a network diagram.

(b) Create a job schedule in total float earliest start sequence, showing the earliest dates within which each job can be done and the total float of each job.

(c) What would be the effect on the estimated completion date of the application if, by date 15, the preliminary survey and the system investigation had been finished, the system design had five weeks of outstanding work, the file design had one week, and the remaining delivery period had dropped to three weeks?

Activity Code	Dates				Floats			
	ES	EF	LS	LF	TF	FFE	FFL	IF
*138	42	43	42	43	0	0	0	0
*140	43	51	43	51	0	0	0	0
*141	51	58	51	58	0	0	0	0
*145	58	60	58	60	0	0	0	0
*146	60	62	60	62	0	0	0	0
*147	62	63	62	63	0	0	0	0
142	51	54	55	58	4	4	4	4
110	42	52	48	58	6	0	6	0
109	52	53	58	59	6	0	0	0
128	53	54	59	60	6	6	0	0
143	43	47	53	57	10	0	10	0
144	47	48	57	58	10	10	0	0
126	42	43	55	56	13	0	13	0
127	43	47	56	60	13	13	0	0
139	43	47	58	62	15	15	2	2
121	42	44	61	63	19	19	19	19

Fig. 13.5 Activity dates and floats for network in Fig. 13.4

Solution to Problem 2. (a) See Fig. 13.6. (b) Job schedule is as follows:

Job	Earliest Start Date	Earliest Finish Date	Total Float
Preliminary survey	0	3	0
System investigation	3	9	0
System design	9	17	0
Write programs C	17	33	0
Test programs C	33	37	0
System test	37	40	0
Write programs A	17	29	5
Test programs A	29	32	5
Write programs B	17	25	10
Test programs B	25	27	10
File design	9	13	15
User-staff training	13	14	15
File creation	14	19	15
Await equipment	3	16	16
Install equipment	16	17	16

(c) By redrawing the network diagram and recalculating the earliest and latest dates of the events, it can be seen that the estimated completion date of the project is delayed by three weeks; i.e., date 43. See Fig. 13.7.

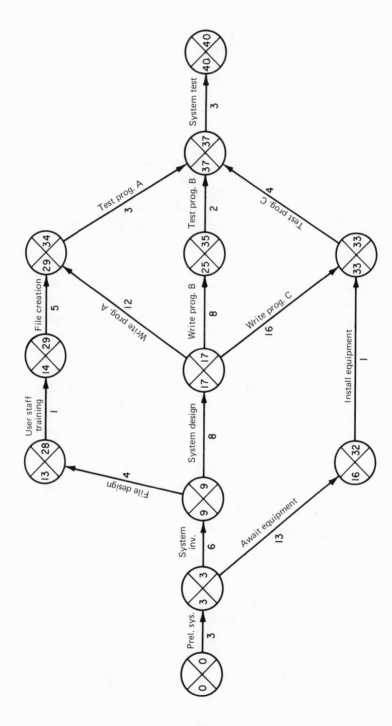

Fig. 13.6 Initial network diagram of Problem 2

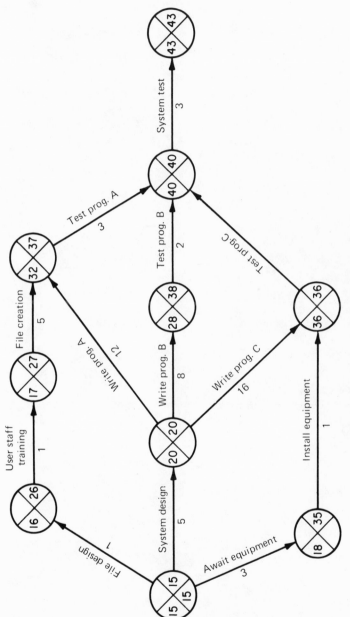

Fig. 13.7 Network diagram of Problem 2 at date 15

285

APPENDIX I

DATA PERTAINING TO THE STATES OF THE UNITED STATES

STATE	AREA, SQUARE MILES*	POPULA- TION, THOU- SANDS**	CAPITAL	PRODUCE
ALABAMA	51,609	3,444	Montgomery	Minerals, cotton, cereals, sugar
ALASKA	586,412	302	Juneau	Timber, furs, fish, metals
ARIZONA	113,909	1,772	Phoenix	Meat, metals, cotton
ARKANSAS	53,104	1,923	Little Rock	Minerals, oil, cotton
CALIFORNIA	158,693	19,953	Los Angeles	Minerals, oil, metals, fruit
COLORADO	104,247	2,207	Denver	Metals, coal, cereals
CONNECTICUT	5,009	3,032	Hartford	Cereals, fish, iron, tobacco, manu- facturing
DELAWARE	2,057	548	Dover	Fruit, cereals, manu- facturing
DISTRICT OF COLUMBIA	67	757	—	Manufacturing
FLORIDA	58,560	6,789	Tallahassee	Fruit, tobacco, sugar, cotton, minerals
GEORGIA	58,876	4,590	Atlanta	Timber, cotton, tobacco, cereals
HAWAII	6,450	770	Honolulu	Fruit, sugar, coffee, hides
IDAHO	83,557	713	Boise City	Minerals, metals
ILLINOIS	56,400	11,114	Springfield	Meat, iron, furs, manufacturing
INDIANA	36,291	5,194	Indianapolis	Meat, manufacturing

287

STATE	AREA, SQUARE MILES*	POPULA-TION, THOU-SANDS**	CAPITAL	PRODUCE
IOWA	56,290	2,825	Des Moines	Cereals, coal, minerals, dairying
KANSAS	82,264	2,249	Topeka	Cereals, dairying, meat, coal, oil
KENTUCKY	40,395	3,219	Frankfort	Tobacco, iron, manufacturing
LOUISIANA	48,523	3,643	Baton Rouge	Timber, sugar, cotton, minerals, tobacco
MAINE	33,215	994	Augusta	Fish, cotton, manufacturing, timber
MARYLAND	10,577	3,922	Annapolis	Coal, minerals, meat, manufacturing
MASSACHUSETTS	8,257	5,689	Boston	Fish, manufacturing
MICHIGAN	58,216	8,875	Lansing	Minerals, manufacturing
MINNESOTA	84,068	3,805	Saint Paul	Timber, meat, cereals
MISSISSIPPI	47,716	2,217	Jackson	Cotton, tobacco, cereals
MISSOURI	69,686	4,677	Jefferson City	Coal, cereals, iron, meat
MONTANA	147,138	694	Helena	Metals, iron, dairying
NEBRASKA	77,227	1,484	Lincoln	Cereals, meat, minerals, fruit, sugar
NEVADA	110,540	489	Carson City	Minerals, metals
NEW HAMPSHIRE	9,304	738	Concord	Fruit, timber, manufacturing
NEW JERSEY	7,836	7,168	Trenton	Cereals, meat, manufacturing
NEW MEXICO	121,666	1,016	Santa Fe	Coal, metals, meat
NEW YORK	49,576	18,191	Albany	Iron, timber, minerals, manufacturing
NORTH CAROLINA	52,586	5,082	Raleigh	Cereals, cotton, tobacco
NORTH DAKOTA	70,665	618	Bismarck	Cereals, coal, meat
OHIO	41,222	10,652	Columbus	Cereals, meat, coal, iron, timber, manufacturing
OKLAHOMA	69,919	2,559	Oklahoma City	Cereals, fruit, cotton, coal, meat
OREGON	96,981	2,091	Salem	Cereals, fruit, metals, coal, fish
PENNSYLVANIA	45,333	11,793	Harrisburg	Minerals, coal, iron, cereals, manufacturing
RHODE ISLAND	1,214	950	Providence	Manufacturing

STATE	AREA, SQUARE MILES*	POPULATION, THOUSANDS**	CAPITAL	PRODUCE
SOUTH CAROLINA	31,055	2,591	Columbia	Cereals, cotton, tobacco
SOUTH DAKOTA	77,047	666	Pierre	Metals, timber, dairying
TENNESSEE	42,244	3,294	Nashville	Cotton, iron, cereals
TEXAS	267,338	11,197	Austin	Cotton, cereals, fruit, coal, oil
UTAH	84,916	1,059	Salt Lake City	Cereals, meat, metals, coal, fruit
VERMONT	9,609	445	Montpelier	Timber, dairying, fruit, meat
VIRGINIA	40,817	4,648	Richmond	Tobacco, cereals
WASHINGTON	68,192	3,409	Olympia	Coal, iron, minerals, timber
WEST VIRGINIA	24,181	1,744	Charleston	Coal, cereals, tobacco
WISCONSIN	56,154	4,418	Madison	Meat, timber, cereals, minerals
WYOMING	97,914	332	Cheyenne	Meat, coal, minerals

*Statistical Abstract of the United States, 1972, 93d ed., U.S. Bureau of the Census.
**Whitaker, *Almanack*, London, 1972.

APPENDIX II

EXAMPLE OF A
COMPUTER RUN SPECIFICATION

This example is the specification of run K12 in Fig. 10.5, and is intended for a programmer who is reasonably experienced and familiar with the data processing department's standards.

General Description of Run K2

This run reads the invoice tape K1 in product code sequence, matches it against the product master cost file T2, also in product code sequence, and writes a new tape K2 (expanded invoice tape). At the same time, unmatched product codes from K1, discrepancies, and control totals are printed.

Volumes of Data

Tape K1—5000 to 7000 records per weekly run.
Tape K2—5000 to 7000 records per weekly run.

The two tapes each have one record per invoice item, unmatched records being omitted from K2. There may be any number of records (including zero) for a given product code.

File T2—600 records approximately, one per product.
Document K22—100 print lines approximately.

Layout of Files

Tape K1—Invoice Tape

FIELD	PICTURE	BYTE POSITIONS IN RECORD
Product code	99999	1–3
Invoice number	A9999	4–6
Area	99	7
Class of trade	99	8
Customer number	999	9–10
Quantity	9999	11–12
Product value, $	999.99	13–15
Discount value, $	999.99	16–18
Week number	99	19
Year	99	20

File T2—Product Master Cost File

FIELD	PICTURE	BYTE POSITIONS
Product code	99999	1–3
Standard labor price, ¢	9999.99	4–6
Standard material price, ¢	9999.99	7–9
Maximum discount, %	99.999	10–12
Selling price, $	9.99	13–14
Description	A(20)	15–34

Tape K2—Expanded Invoice Tape

FIELD	PICTURE	BYTE POSITIONS
Area	99	1
Class of trade	99	2
Customer number	999	3–4
Product code	99999	5–7
Standard labor value, $	999.999	8–10
Standard material value, $	999.999	11–13
Product value, $	999.99	14–16
Calculated value, $	999.99	17–19
Discount value, $	999.99	20–22
Week number	99	23
Year	99	24
Quantity	9999	25–26

Magnetic Tape Block Lengths

Tape K1	25 records of 20 bytes = 500 bytes
Tape K2	19 records of 26 bytes = 494 bytes
File T2	14 records of 34 bytes = 476 bytes

Processing

1. Check that the week and year are as parameters; if not, stop run and display as per operating instructions.
2. Read K1 record and match its product code against T2. If matched, add 1 to control count H. If unmatched, add 1 to control count J, print as per output layout line A, and ignore K1 record.
3. Extract selling price from T2; multiply by quantity equals calculated value.
4. Compare calculated value with product value from K1. If unequal, add 1 to control count K, and print as per output layout line B.
5. Add calculated value to control total D. Add product value to control total E. Add quantity to control total G.
6. Extract standard costs from T2, multiply by quantity, and insert results into K2.
   Standard labor value = standard labor price X quantity.
   Standard material value = standard material price X quantity.
   No rounding is necessary.
7. Multiply: calculated value X maximum discount percent = maximum discount value.
8. Compare maximum discount value with discount value from K1.
   If discount value is greater than maximum, add 1 to control count L, and print as per output layout line C.
9. At end of K1, calculate: total value discrepancy = total calculated value less total product value. Print as per output line F, with * if positive, – if negative.

Output Layouts

Since this is an internal document to be printed on blank stationery, the exact print positions are not important, but the layout is as below.

Number of copies: 2
Spacing: double
Lines per sheet: 30
Sheets numbered consecutively.

	PRODUCT CODE	INVOICE NO.	CAL- CULATED VALUE	PRODUCT VALUE	MAX- MUM DIS- COUNT	DISCOUNT VALUE	COMMENT
A	44697	E8093					Missing product code on T2
B	48560	E8562	176.70	175.06			Value discrepancy
C	51233	E8411			37.34	56.01	Discount above maximum

D	Total calculated value	177,183.76
E	Total product value	176,631.62
F	Total value discrepancy	552.14*
G	Total quantity	131,265
H	Matched records count	6,318
J	Unmatched records count	46
K	Value discrepancy count	28
L	Discount above maximum count	16

INDEX

295